Tu Youyou

and the Discovery of Artemisinin

2015 Nobel Laureate in Physiology or Medicine

Tu Youyou

and the Discovery of Artemisinin

2015 Nobel Laureate in Physiology or Medicine

Yi Rao
Daqing Zhang
Runhong Li
Peking University, China

World Scientific

NEW JERSEY • LONDON • SINGAPORE • BEIJING • SHANGHAI • HONG KONG • TAIPEI • CHENNAI • TOKYO

Published by

World Scientific Publishing Co. Pte. Ltd.
5 Toh Tuck Link, Singapore 596224
USA office: 27 Warren Street, Suite 401-402, Hackensack, NJ 07601
UK office: 57 Shelton Street, Covent Garden, London WC2H 9HE

Library of Congress Cataloging-in-Publication Data
Names: Rao, Yi, 1962– editor. | Zhang, Daqing, 1959– editor. | Li, Runhong, editor.
Title: Tu Youyou and the discovery of artemisinin : 2015 Nobel laureate in physiology or medicine / editors, Yi Rao, Daqing Zhang, Runhong Li.
Description: New Jersey : World Scientific, 2016. | Text in English, with some Chinese.
Identifiers: LCCN 2016019164| ISBN 9789813109889 (hardcover : alk. paper) | ISBN 9789813109896 (pbk. : alk. paper)
Subjects: | MESH: Tu, Youyou, 1930– | Artemisinins--history | Antimalarials--history | Drug Discovery--history | Nobel Prize | History, 21st Century | China
Classification: LCC RS420 | NLM QV 256 | DDC 615.1/9--dc23
LC record available at https://lccn.loc.gov/2016019164

British Library Cataloguing-in-Publication Data
A catalogue record for this book is available from the British Library.

This work is a translation of《呦呦有蒿: 屠呦呦与青蒿素》/ 饶毅, 张大庆, 黎润红等著, Beijing: China Science and Technology Press, 2015.
ISBN 978-7-5046-6996-4.

Translators: Min-Jun Neo and Kok-Hsien Tan
Copy Editor: Sharon Khoo

Tu Youyou

Pharmacist

Born: 30 December 1930, Ningbo, Zhejiang Province

Graduated from the Department of Pharmacy, Beijing Medical College (now Peking University Health Science Center) in 1955

Life Fellow and Principal Research Scientist in China Academy of Chinese Medical Sciences

Director of Artemisinin Research and Development Center, PhD Supervisor

Winner of the 2011 Lasker-DeBakey Clinical Medical Research Award

Winner of the 2015 Nobel Prize in Physiology or Medicine

Artemisinin is a precious gift from traditional Chinese medicine to mankind. Compared to other practical uses of phytochemistry in drug development, although the history of artemisinin is relatively short, it is not the only fruit of the wisdom of Chinese medicine.

— Tu Youyou

Contents

Authors

Yi Rao was born in Nanchang, Jiangxi.

He is a tenured professor of Peking University and Senior Research Fellow of the National Institute of Biological Sciences, Beijing.

He obtained his doctorate in neuroscience from the University of California at San Francisco in 1991.

He has served on the faculty of the University of Washington and Northwestern University. Other positions he has held at Peking University include Deputy Director of the Neuroscience Research Institute and Dean of the School of Life Sciences.

He has recently been appointed Director of the Faculty of Science, Peking University. His major research interests are in the molecular and biological mechanisms of neurological development and behaviours.

Daqing Zhang was born in Shashi, Hubei.

He is currently the Director of the School of Medical Humanities and a professor at the History of Medicine Research Center, Peking University.

Other positions he has held include:

Head of the Peking University Health Science Library, Vice Chair of Chinese Society of History of Science and Technology, Vice Chair of the Society of History of Medicine, Vice Chair and Director of the Chinese Society for Dialectics of Nature, Philosophy of Nature, Science and Technology, and Deputy Director of the Ministry of Education Steering Committee on the Curriculum of Medical Humanities and Social Sciences of Medical Colleges.

Runhong Li was born in Lichuan, Jiangxi.

She is an assistant researcher at the School of Medical Humanities, Peking University. She obtained a postgraduate degree in medicine at Peking University in 2011. Her main research focus is on the history of medicine.

Preface

The Deer Bleat "Youyou" While They Feed on Wild Hao

The effect of artemisinin is now widely known. Fast-acting artemisinin can be used as first line treatment against malaria, and as an alternative drug when there is resistance to common drugs such as chloroquine. Of course, artemisinin also has its limitations, and cannot be a replacement for all the other anti-malarial drugs. However, it has cured many people. Artemisinin is a drug with a structure completely different from other anti-malarial drugs. Until now, China is still seeking a derivative of artemisinin which is more effective and less likely to give rise to drug resistance. The mechanism of artemisinin is not completely understood and is a question that should be explored further.

Since early 2000, the discovery of artemisinin and Tu Youyou's role has gained prominence.

In 1969, senior scientists were set aside and were unable to participate scientific research. During that period, Tu Youyou, a research apprentice in the department of Chinese traditional medicine of the Academy of Traditional Chinese Medicine (ATCM) was assigned to work on "Project 523".

Project 523 involved several parts such as reproducing western medicine or creating derivatives, seeking antimalarial drugs in Chinese traditional medicine, creating mosquito repellants etc. In the Chinese medicine section, different research groups tested various Chinese medicines, including *Dichroa febrifuga* which had a strong curative effect but with more side effects. In the 1940s, Zhao Changshao and other scientists initiated pioneer research in *Dichroa febrifuga*. In 1943, they reported that a crude extract from *Dichroa febrifuga* could treat malaria. Following that, in 1945, Zhao reported that three alkaloids extracted from *Dichroa febrifuga* were effective in malaria-infected chickens in 1945. In 1946, he reported that dichroine b was effective on malaria-infected chickens. In 1948 Zhao Changshao reported that extracts from *Dichroa febrifuga*, dichroine g, dichroine b, dichroidine and quinazolone, were effective against malaria as well, and analysed the chemical formula of these alkaloids. Project 523 reconsidered *Dichroa febrifuga*, but faced same problem: although *Dichroa febrifuga* had a strong curative effect, it caused nausea and was not suitable for widespread use. However, the research into *Dichroa febrifuga* followed almost the same route as for *Artemisia annua* (*A. annua*) and artemisinin.

A. annua not only appeared in ancient Chinese medical texts but was also used by the Chinese people in 1950s and 1960s. Yu Yagang, a member in Tu Youyou's research group, made a list of all possible Chinese medicines that might be effective against malaria, resulting in a list including 808 Chinese medicine prescriptions, including *Aconitum*, *Prunus mume*, turtle shell and *A. annua*. The Academy of Military Medical Sciences filtered out almost a hundred prescriptions by testing them on malaria-infected mice and discovered that extracts from *A. annua* had 60–80% effectiveness against malaria, but the effectiveness was unstable. The list provided by Tu Youyou to her research group included several Chinese medicines derived from minerals such as litharge, realgar, sulfur, copperas and cinnabar; from animals such as pill bug, earthworm, pillis ophidiae, pangolin and chorion ovi; and from plants such as *Cortex lycii*, *Radix kansui*, *Hemerocallis citrina Baroni*, *Eleocharis dulcis*, *Brucea javanica*, *A. annua* and *Verbena officinalis* L. In early 1971, Yu Yagang was transferred from the anti-malaria research group to the bronchitis research group. Tu Youyou's research group also noticed the effect of *A. annua*, but the water decoction of *A. annua* was ineffective and the effectiveness of extracts using 95% ethanol were only 30–40% effective. It should be pointed out that unreliable records referring to water decoctions of *A. annua* for malaria treatment in ancient books also hampered to some extent the discovery of the real function of the extract.

In the second half of 1971, Tu Youyou proposed using ether to extract *A. annua* and this increased the antimalarial effectiveness to 95–100%. This method was the key to the discovery of the effectiveness of crude extracts of *A. annua*. In March 1972, Tu Youyou's report of this research result at a Project 523 meeting in Nanjing attracted some attention, but it was not regarded as an important result. The summing up at the conclusion of the meeting suggested that researchers should proceed to "determine the chemical structure of *Artabotrys hexapetalus* as soon as possible and continue the research on its synthesis; to further confirm the effectiveness of the effective monomer *Agrimonia pilosa* and clarify its chemical structure; to test the clinical efficacy of *A. annua* and *Ailanthus altissima* while speeding up the separation and extraction of their effective monomers."

Thereafter, Tu Youyou's research group focused on *A. annua*. After Ni Muyun extracted the active compound in *A. annua* and Zhong Yurong obtained crystals of artemisinin II (later known as artemisinin), Tu Youyou presented the chemical formula of artemisinin II at the *A. annua* Research Forum held by the ATCM in February 1974. The research group first tested the anti-malarial effects of *A. annua* ether neutral extract (an effective anti-malarial compound in the form of a black gel), then later obtained artemisinin (an effective anti-malarial monomer in the form of white needle-shaped crystals).

Tu Youyou's research group collaborated with other research groups, especially with the Shanghai Institute of Organic Chemistry and the Institute of Biophysics of the Chinese Academy of Sciences (CAS), to analyse the artemisinin molecule and its structure, discovering that artemisinin was a new type of sesquiterpene lactone. Inspired by the research result from Tu Youyou's group, in 1972, the Shangdong Institute of Parasitic Diseases collaborated with the Shangdong Academy of Chinese Medicine perform extraction from *A. annua*, while the Yunnan Institute of Materia Medica worked independently. The Shangdong Academy of Chinese Medicine and the Yunnan Institute of Materia Medica each obtained an effective anti-malarial monomer, naming it "arteannuin" (Shandong) and "artemisinin" (Yunnan) respectively. In early 1974, artemisinin (*qinghaosu*) from Beijing, arteannuin (*huanghuahaosu*) from Shandong and artemisinin (*huanghuasu*) from Yunnan were confirmed to be the same drug.

Based on the analysis of the history of the discovery of artemisinin, despite considerable debate, the following issues are undeniable:

1. Tu Youyou proposed the method of extracting artemisinin with ether that was the key to discovering the anti-malarial effect of artemisinin and advancing the research on artemisinin;
2. Zhong Yurong, who performed the actual separation and purification of artemisinin, was a member of Tu Youyou's research group;
3. Other research groups who obtained artemisinin by extraction after hearing of the report from Tu Youyou's research group at a meeting obtained the purified molecule after Zhong Yurong.

There are many accounts relating to the discovery of artemisinin. The discovery of a drug comprises many stages: besides determining the effectiveness of a crude extract, other stages include purification, pharmacology, structure and clinical practice etc. Tu Youyou's achievement was based on predecessors' work, and members of her research group made valuable contributions. Important contributions from other research groups and scientists should not be overlooked either. For example, the ATCM learned extraction techniques from Yunnan and Shandong. Experimenting with the crystals they extracted, the ATCM discovered that the results of clinical trials were not satisfactory and the extracts had toxic side effects. Meanwhile, Li Guoqiao from the Guangzhou University of Chinese Medicine verified that crystals extracted by Luo Zeyuan and his colleagues from the Yunnan Institute of Materia Medica were effective against *Plasmodium falciparum* malaria, especially cerebral malaria. The wide-spread use of artemisinin derivatives such as artemether and artesunate was result of much research by Li Ying of the Shanghai Institute of Materia Medicine, CAS, and Liu Xu from the Guilin Pharmaceutical Factory.

Looking at the research done by that generation of scientists, it can be seen that they remained unknown to the public even though their drugs saved many lives. Many of the relevant documents are buried in unknown academic journals and internal meeting records which are not easy to access.

With material gleaned from the original Chinese papers, documents and journals, this book presents the history of these discoveries. At the same time, we are mindful that there may be other treasures from ancient and contemporary Chinese documents and medical practices still waiting to be discovered.

Chapter 1
Project 523 and Artemisinin

This is an international honor for the Chinese people, Chinese science and Chinese medicine.

— Tu Youyou

The Mandate for Project 523

The project origins

In the early 1960s, the chloroquine-resistant *Plasmodium falciparum* (*P. falciparum*) strain of malaria began to emerge in many parts of the world, especially in Southeast Asia. As the Vietnam War escalated, chloroquine-resistant *P. falciparum* ran rampant, endangering the lives of Vietnamese soldiers and citizens. In 1964, Chairman Mao Zedong had a meeting with a Vietnamese communist official who raised the issue of the serious malaria epidemic in Vietnam and requested help to deal with it. Chairman Mao replied, "Solving your problem would be solving our problem as well". After that, the General Logistics Department (GLD) issued an order to the Academy of Military Medical Sciences (AMMS) and the Second Military Medical University (SMMU) to develop a long-acting and effective anti-malarial drug. These two units were to carry out the research concurrently. Even before 1967, the AMMS, SMMU and the Institutes of Military Medical Sciences in Guangzhou, Kunming and Nanjing had already been researching various aspects of emergency war assistance, including the prevention and treatment of malaria. AMMS had already recommended that anti-malarial medication be administered to the troops training and fighting in regions where malaria was endemic. However, the common anti-malarial drugs were short-acting and had to be taken regularly. When used on a large scale, there were cases of missed dosages or unwillingness to consume the drug which affected their effectiveness in malaria control. Therefore, the AMMS had begun a search for more effective and longer-acting anti-malaria drugs. For instance, the Sixth Institute of the AMMS had synthesised CI-501, a long-acting anti-malaria drug, and carried out extensive research on *Dichroa febrifuga*, a traditional Chinese herb. The Fifth Institute of the AMMS also carried out a large number of animal studies to test the efficacy of CI-501 on malaria-infected mice and chickens.

From May to August 1966, the AMMS deployed a large number of personnel to Vietnam to survey the overall health, incidence of disease outbreaks, and the prevention and treatment of disease among the Chinese troops stationed in Vietnam with a focus on the incidence, prevention and treatment of malaria. According to Ruan Guozhang, the North Vietnam Military Health representative, malaria was the main infectious disease affecting the troops. Malaria was widespread in the Vietnam People's Army, and the incidence of malaria was 100% among the South Vietnamese troops as well as in the US Army's First Division. Most of the common anti-malarial drugs had lost their effectiveness against the drug-resistant *P. falciparum* strain and there was an urgent need to develop new anti-malarial drugs. Since the research capability of the military alone was insufficient, a decision was subsequently made to allow the GLD to work together with the State Science and Technology Commission (STC) to set up a wider collaboration. After discussions, preparations and authorisation by the relevant departments, on 4 May 1967, the STC announced the convening of an anti-malaria drug research collaboration work meeting. On 18 May, the national leadership team for anti-malaria research held a meeting in Beijing. From 23 to 30 May 1967, members of the STC and the GLD held meetings with relevant ministries and commissions, work units directly under the Central Military Commission (CMC), as well as leaders from affected provinces, cities, autonomous regions and military regions. The meeting approved a three-year research plan drawn up by the AMMS. This was a classified as an emergency military project involving the Vietnam War. For confidentiality, the meeting date was abbreviated and the research plan was given the codename "Project 523". Other related names were "Project 523 Leadership Group" and "Project 523 Office".

The meeting preparations

From 23 to 30 May 1967, the STC and GLD convened the Anti-Malaria Drug Research Work Meeting in Beijing. There were a total of 88 representatives from 37 participating work units. These were relevant specialist departments as well as units involved in the production, research and testing of anti-malaria drugs. As the Cultural Revolution was then at its height, the complete record of the meeting is unavailable as the records of the meeting have not been preserved. Over time, the surviving participants of the meeting grow fewer, and therefore information about the meeting preparations and proceedings based on the recollections of these participants is extremely valuable. We visited some of the leaders and researchers who participated in Project 523. Among them was one of the pioneer members, Wu Zilin, former director of the Science and Technology Planning Department of AMMS. Other members we interviewed were Chen Haifeng, former director of the Science and Technology Division of the Ministry of Health (MOH), and Fu Liangshu, chairman of the Yunnan Project 523 Office.

(1) Planning for the meeting

Wu: Around June 1966, I returned from conducting a nuclear experiment. Gui Shaozhong, the director of AMMS, asked me to attend a meeting with the GLD. Chen Pang, chief of staff of the GLD told us of an order from Chairman Mao. In brief, the order stated: "The Southern forces in Vietnam are suffering from malaria. Their combat ability has been affected, jeopardising the survival of the whole country. The military research team must solve find a way to prevent and treat malaria." Even before Chairman Mao's order, the AMMS had been engaged in research into the prevention and treatment of malaria. Drug and chemical synthesis were the forte of the Fifth and Sixth Institutes of the AMMS and they had developed the malaria prevention tablets No. 1 and No. 2. At that time, the research was mainly undertaken by the military, while the STC and MOH were not yet involved. Since I considered that the AMMS did not have the capability to carry this research out alone, after returning from the meeting, I spoke to the director of the Science and Technology Planning Department (who had not attended the meeting) and began to plan for a wider collaboration.

At that time, military research projects usually remained at the planning stage until an order to proceed was received. One of the directors of the AMMS, He Chen, had high language skills. When we were appointed to draft the initial research plan, it was mainly he who spoke while we took minutes. He suggested that the plan deal with two main areas — directions, and tasks. The main directions were to investigate, experiment, eradicate, prevent, and treat. Following on from these directions, we began planning the collaboration. Our aim was to combine science, education and research cooperation within the military as well as with civilian work units, and to include clinical trials, drug manufacturing and field experiments. These ideas were reflected in the plan and served as guidelines as we drafted subsequent plans and implementation procedures for participating work units. After the meeting in 1966, the AMMS established a team of five, consisting of director Gui Shaozhong, assistant director Peng Fangfu, Xu Nianzi of the Sixth Institute of the AMMS, Bai Bingqiu of the Fifth Institute of the AMMS and Liu Demao. However, after the onset of the Cultural Revolution, most members of the team were unable to participate in the meeting on 23 May 1967. Only Peng Fangfu attended. The collaboration began after this meeting. Subsequently, Liu Demao was placed in charge. After discussion, I was appointed the team leader, with assistant leaders Zhou Yanchong, who was mainly in charge of the technological aspects, and Zhou Yiqing.

During the turbulent years of the Cultural Revolution, many work units had ceased operating. To enlist the participation of these work units, I went personally to inform them about the order. I first approached the internal work units of the military. There was no problem obtaining their collaboration as the military districts would unquestioningly carry out orders issued by the headquarters. In those days, the SMMU was one of the more important work units due to their strong research capabilities. I first approached the SMMU which had already done much research on piperaquine. At that time, several military universities had been seized by the Red Guards, and the SMMU had already been occupied. The directors and principal were stripped of their powers and were being watched by the Red Guards. I was at a loss as I did not have any connections with anyone in the Red Guards there. With no supporting documents, I would have to communicate Chairman Mao's orders verbally to them.

I received help from a revolutionary group in AMMS which had contacts with the Red Guards. They wrote a letter to the head of the SMMU Red Guards, a lady surnamed Qian, who immediately agreed to carry out Chairman Mao's orders. She brought me to meet Wang, the director of the SMMU Science and Technology Department, and some other experts and professors who were all willing to participate in the research rather than the revolutionary activities. Thus it was settled. I then proceeded to the Ministry of Chemical Industry (MCI). At the time the factories supported the revolutionaries. At the top of the hierarchy was the STC, directed by Nie Rongzhen. However, the STC had been stripped of its powers and he did not have any authority. At the time, a lady in STC named Zhang Ben was very supportive of the collaborative research. A few directors in STC were quite supportive, especially director Tian, who was involved in drug research. Qian Xinzhong from the MOH had been stripped of power, and Chen Haifeng and others were all being watched by the revolutionaries. Beijing Medical University did not participate. Several Chinese physicians and local work units participated as well as a few pharmaceutical factories which were incorporated by members of the STC and the Commission for Science, Technology and Industry for National Defense (COSTIND). The local work units were simply told that it was an order from Chairman Mao. Even so, this was the only issue that I was investigated for during the Cultural Revolution as they felt that my orders were unclear.

Chen: On 18 April 1967, Bureau 10 of the STC convened the Anti-Malaria Drug Research Collaboration Meeting. I am unclear about the details, but this was before Project 523. I was assigned to the project on 18 May, and a meeting was held to establish the anti-malaria research leadership team. By the time I was appointed, the project had already been approved by Chairman Mao and Prime Minister Zhou. After their approval, the representatives of the military had gone to the STC, GLD, Chinese Academy of Sciences (CAS), MCI and MOH to discuss the formation of a Leadership Group to for anti-malaria research.

(2) The meeting itself

Wu: The meeting itself was very chaotic. Originally, it was supposed to be held at the GLD. However, as the SMMU Red Guards had decided to hold revolutionary activities there, Qiu Huizuo, People's Liberation Army (PLA) Deputy Chief of Staff and head of the GLD decided to hold the meeting at the Beijing Hotel instead. Liu Demao, Zhou Tingchong and I went to the hotel together. Since there were so many rooms at the hotel, one could choose any room to stay in. At that time, the new hotel building was still under construction and all the rooms available were old. More than a hundred people attended the meeting. The actual meeting lasted only two to three days, while the rest of the time was used for discussion. All in all, we were there for about a week. During the meeting, I presented the proposal while Liu Demao took charge of organising the meeting. Assistant Director Peng Fangfu attended the meeting, but since he opposed our approach, he did not speak; he just signed the documents. So basically, it was my job to speak. The meeting was very interesting in that before anyone could make a comment, they had to shout out a slogan. The old Red Army men present were unused to it, and so were we, but we had no choice. Those who attended the meeting were from the revolutionary faction as those from the conservative faction had been stripped of their power and were unable to attend the meeting. After the proposal had been heard, the participants split into smaller groups for group discussions, and the meeting went smoothly. Subsequently, meetings were held by the individual groups. Zhou Keding attended the meeting as my personal assistant.

The groups were assigned according to their responsibilities and work units. There were no further instructions received after the meeting, and we made the final decisions. The formal documents were submitted to the GLD and STC after the meeting. I cannot recall the exact procedures clearly now.

Chen: The first meeting was held at the Beijing Hotel from 23 to 30 May and the proposal was officially presented. The proposal at that time was in rather rough form, and the responsibilities were first assigned to kick-start the project. This emergency war preparation project was highly confidential. Other than the Leadership Groups, Deputy Minister Yang Dingcheng of the PLA's department of health was also present. He introduced the project which was to find the fastest-acting and most effective solution for malaria, especially against *P. falciparum*. After his introduction, everyone proceeded to discuss the project and a preliminary plan was drawn up. Since the meeting was held on 23 May, the plan was coded "523 Office" for confidentiality. The military named it "Military 523 Office", and the local districts named it "Regional 523 Office". In the provinces and cities, the regional offices and the military worked together but it was rather chaotic. The military, led by the AMMS, generally took the lead. Its director then was Zhang Jianfang, and another important researcher was Bai Bingqiu from AMMS Fifth Institute. The microbiology and epidemiology research institutes, led by Zhou Yiqing and others, focused more on the field work of malaria prevention and treatment. The Fifth Institute focused on malaria prevention and was headed by Bai Bingqiu. A group of personnel from Fifth Institute, including Tian Xin and many others had been engaged in field work before. At that time, Liu Jichen, a research assistant at the research department of the MOH was the backbone of the research. Zhang Jianfang was the leader, together with Bai Bingqiu, Liu Jichen and Zhou Keding. The late Zhou Keding was very strong in his field of expertise and would have been the one with the greatest understanding of the whole situation. During the period when the 523 Office was formed, the two opposing political factions were already at loggerheads and the leaders of the cities and provinces were all in a state of partial or complete paralysis. When the two opposing factions were fighting, it was very difficult to contact anyone. At that time, the meeting was attended by representatives from both factions. Chairman Mao had authorised the meeting, and Prime Minister Zhou undertook to get hold of the members. Whenever Prime Minister Zhou needed to do anything, he would have to send a written recommendation letter from the State Council as the recommendation letter by the MOH was ineffective. Whenever we went to the provinces, we would have to bring along the recommendation letter from the State Council, otherwise we would not be able to get hold of anyone. At the provincial level, the State Council recommendation letter was effective as everyone had to obey the order. We used the letter to show that we were executing war preparations so no-one could hinder our mission, and everyone had to provide full support. We refused to get involved in revolutionary activities or factional fighting. We were there to promote production, and that was the basis for our work. We referred factional issues to the local Revolutionary Committee. Our aim was to promote production and to engage in the secret "523 Research". We held a few trump cards: authorisation from the central government and a recommendation letter the State Council stating that we were engaged in secret war preparation. Usually, the leaders of the both factions would allow their people to participate after seeing my trump cards. Within each faction, there were some who promoted production. Although it had a difficult start, the research gradually got under way according to our plan. At the time, our responsibility was to follow the plan without alterations, but once a problem was identified, we would handle it.

Fu: In May 1967, we were told to prepare for an Anti-Malaria Drug Research Work Meeting. The project was codenamed "523" for confidentiality such as "523 Office", "523 Leadership Group", "Project 523", "523 Meeting" etc. The Anti-Malaria Drug Research Work Meeting was led by the GLD. Other participating units were the MOH, STC, CAS, Academy of Medical Science, AMMS etc. The participants from AMMS were mainly from the Fifth Institute, as earlier malaria research had been mainly carried out by them. At that point in time, AMMS had already formed a group within their planning department and chosen some people to take charge and draft the plan. Deputy head of the Kunming Military District health department, Yi Zhimei, Deputy Director of the AMMS, He Bin (from Taiwan), and I attended the meeting. I understand that the civilians from our district were all from the revolutionary faction. The attendees from the army and the Fifth Institute also included those from the revolutionary faction. Besides our military district, there were participants from the Guangzhou military district, which included Hainan. Participants also came from other military medical universities, and the National Institute of Parasitic Diseases. The Guangzhou military district and our Kunming military district were the two main participants. Participants were mainly from regions where malaria was endemic, and the National Institute of Parasitic Diseases was there to help in the research. The chief aim of the meeting was to assign responsibilities.

The meeting lasted for under a week, probably four to five days. I recall that the participating leaders were AMMS Director Peng Fangfu, head of the Fifth Institute Bai Bingqiu, commissioner Tian Ye from the STC, and Director Chen Haifeng from the Science and Technology Division of the MOH. All of them sat at the back of the meeting, too afraid to make any comments, because as leaders, they were all targets of the revolutionary faction. In fact, the revolutionary faction had been kind to even allow them to attend the meeting. At one point we were told to criticise Liu Shaoqi and Deng Xiaoping. We were in Beijing at that chaotic time, and although the revolutionary faction organised the meeting, it was the directors named above who made the plans and assigned the tasks. Not many people were present, only about a hundred people in a small meeting room, unlike subsequent meetings which were attended by more people. At this meeting, it was considered sufficient that every work unit was represented. The meeting was mainly to assign the tasks and to get the work started.

From the participants' accounts, it was clear that the meeting was held at a turbulent time. The meeting itself was rather disorganised too. Nevertheless, despite the ongoing Cultural Revolution, the meeting convened with the approval of the highest leader. This showed the importance of this mission.

The mission unfolds

The Anti-Malaria Drug Research Work Meeting ended on 30 May 1967. Not long after that, on 16 June, the STC and GLD issued a joint notice to formally release the summary of the Anti-Malaria Drug Research Work Meeting, the Anti-Malaria Drug Research Work Plan and the arrangements for Project 523. The summary formally defined six departments, the STC, COSTIND, CAS, GLD, MOH and MCI, as the leaders of the project. The participating work units were grouped into four main collaboration teams according to their expertise, namely drug synthesis and screening, traditional Chinese medicine, mosquito repellants, and field prevention and treatment (including drug manufacturing). The teams were also grouped accordingly to locations: the eastern, northern, southwestern and northeastern teams. The urgency and importance of the project were emphasised again.

The Anti-Malaria Drug Research Work Plan gave a detailed description of the specific job scopes. The plan was based on the proposal that the AMMS drafted before the meeting and revised after discussion with the participants at the meeting.

The specific objectives in the plan were to research:

1) drugs to prevent and treat drug-resistant malaria;
2) anti-malaria drugs which were longer-acting; and
3) mosquito repellents.

The detailed description of each task was elaborated in the plan. For example: Within two to three years, to develop three to four different types of new drugs effective against drug-resistant malaria for military use; besides overcoming the drug resistance issue, the drugs were to be long-acting. To develop long-acting topical or oral mosquito repellants. The topical repellant should be effective for at least 24 hours and the oral repellant at least 12 hours, and they had to be safe, convenient and with minimal side effects.

These three objectives were further divided into five research topics (Figure 1):

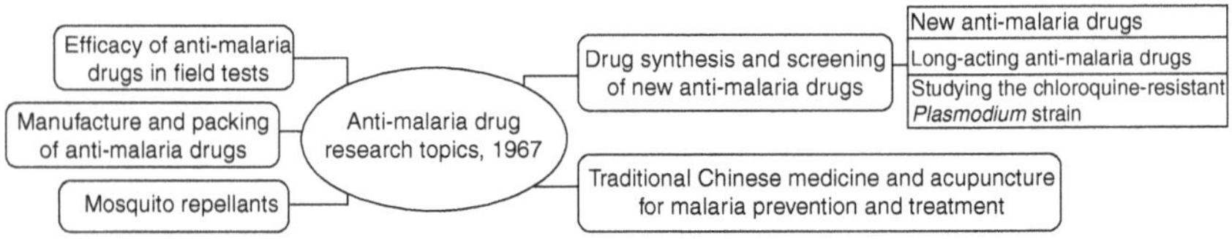

Figure 1: The initial three-year plan and the five research topics

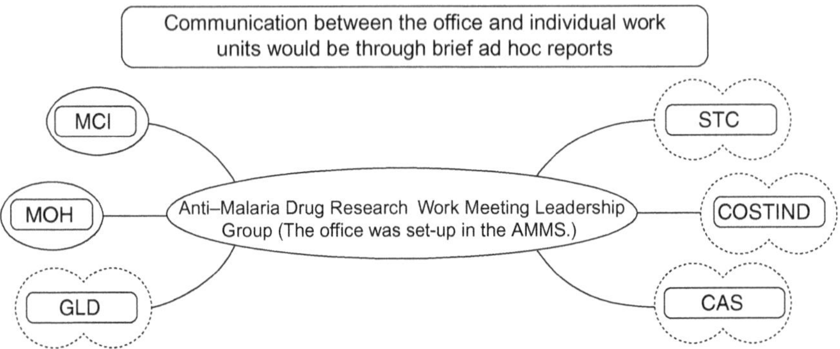

Figure 2: Anti-Malaria Drug Research Work Meeting Leadership Group, 1967

For each subgroup, the details of how the work was to be distributed among participating work units, the main research topics and expected rate of progress were spelled out.

It was felt that the leadership should be strengthened for better implementation of the plan. This was to achieve tighter collaboration and speedier communication of experiences between the different work units. Thus, the plan proposed the following:

1) The Anti-Malaria Drug Research Work Meeting Leadership Group (see Figure 2):
2) Four specialist teams (Figure 3):
3) All organisations responsible for the planning of the tasks were to take the job seriously. These organisations included provincial or city science and technology commissions, military districts, health

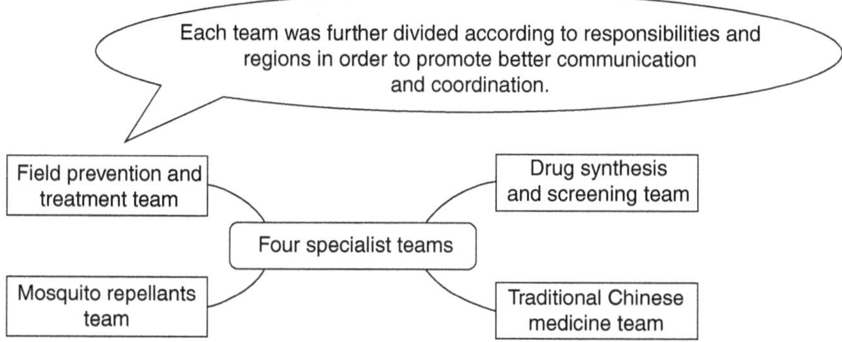

Figure 3: The four specialist teams, 1967

ministries and other work units. Scientific and technical personnel were to work closely with peasants and soldiers. The laboratories were to work closely with those conducting field clinical experiments to enable rapid completion of this emergency war mission.

4) That the execution of this research task should be classified as secret.

Summary execution plan for Project 523 (Figure 4):

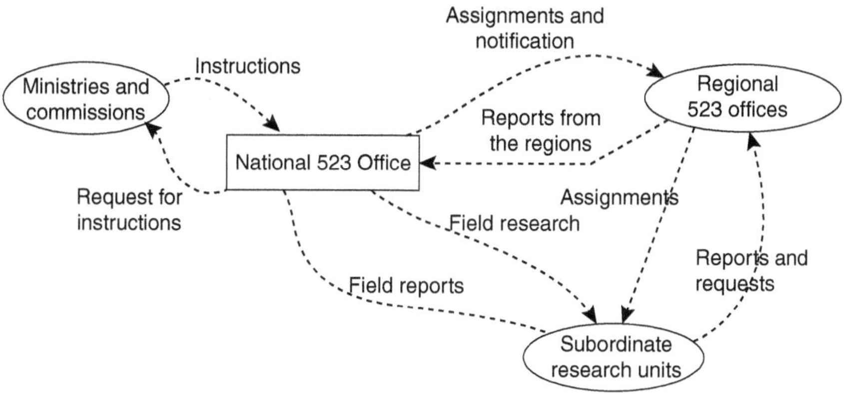

Note: Communication flows between the organisations in the form of various meetings.

Figure 4: Project execution plan

The Organisational Structure of Project 523

The initial years (1967–1971)

(1) Organisational structure

The composition of the Anti-Malaria Drug Research Leadership Group was finalised during the meeting held from 23 to 30 May 1967. The organisation structure was as follows (Figure 5):

The Leadership Group was formed with one representative each from STC, COSTIND, CAS, GLD, MOH and MCI. They reported directly to the STC. The Leadership Group set up an administrative body that was staffed mainly by the AMMS with a representative from each of CAS, Chinese Academy of Medical Sciences (CAMS) and China National Pharmaceutical Industry Corporation (CNPIC). The office was located at the AMMS and was in charge of handling the day-to-day operations of the research collaboration and communicating the results of the research. The team was led in succession by Peng Fangfu, AMMS Vice Principal Major-General Qi Kairen. Bai

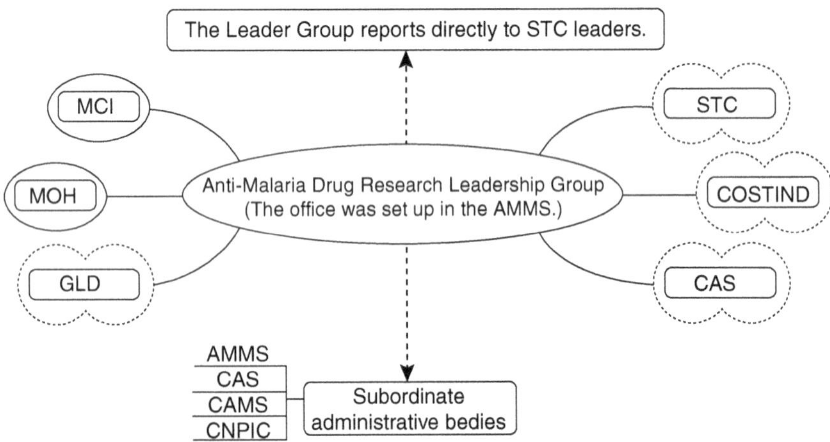

Note: The Leadership Group consisted of one representative from each of the six leadership organisations. The administrative body was mainly staffed by AMMS, with one representative from each of the other three departments.

Figure 5: The organization structure of the Leadership Group

Bingqiu was the office manager and Zhang Jianfang was the assistant manager.

The five topics in the Anti-Malaria Drug Research Work Plan were:

1) drug synthesis and screening;
2) traditional Chinese medicine and acupuncture;
3) mosquito repellants;
4) manufacturing and packaging of drugs; and
5) prevention and treatment in the field.

These five topics were divided between four collaboration teams. The team leader, assistant leader and corresponding tasks were also specified in the plan. The four specialist teams were:

1) Drug synthesis and screening team
   Led by the AMMS and assisted by the Shanghai Institute of Pharmaceutical Industry and the Academy of Traditional Chinese Medicine (ATCM). It was further divided into two regional teams: the Eastern China team and North/Northwest/Southeast team.
2) Traditional Chinese medicine team
   Led by the ATCM and assisted by the Shanghai Research Institute of Acupuncture and AMMS.
3) Mosquito repellants team
   Led by AMMS and assisted by the Shanghai Institute of Pharmaceutical Industry and the Third Military Medical University.
4) Field prevention and treatment team (including drug manufacturing)
   Led by AMMS and the National Institute of Parasitic Diseases, CAMS, and assisted by the Military Medical Institute of the Logistics Department of the Kunming Military District, the Health and Epidemic Prevention Institute of the Logistics Department of the Guangzhou Military District and the health department of the Logistics Department of the Nanjing Military District.

The responsibilities of the leaders of each team were:

1) to survey and assess the situations;
2) to share Chairman Mao's experiences and work experiences from his writings;
3) to coordinate and promote the implementation of the work plan; and
4) to communicate information between the leaders and the workers.

The second Anti-Malaria Research Work Meeting was held in 1968 and the responsibilities of the Leadership Group and each department were spelled out in more concrete terms. Although the general thrust of the research task outlined in the three-year plan agreed upon at the 1967 meeting was mostly left unchanged, the responsibilities of each organisation in the Leadership Group were described in greater detail. In addition, formal rules as to confidentiality were set out in writing.

Leadership Groups at all levels were required to be named "523 Leadership Group", the offices were to be named "523 Office", and the collective name for all specialist groups was "523 Specialist Team". The traditional Chinese medicine team was named "523 First Team", the drug synthesis team was named "523 Second Team", the acupuncture team was named "523 Third Team", and the mosquito repellants team was named "523 Fourth Team". Any drugs that underwent clinical trials or went into production were to be given a code name. Participating research personnel were bound by the secrecy regulations and were to be checked regularly by each work unit.

At that meeting, a decision was made by the Leadership Groups to change the name on the official seal from "Anti-Malaria Drug Research Leadership Group Office" to "Beijing 523 Leadership Group Office".

The amendment of the seal reflected the secrecy of the work. After the 1968 meeting, in addition to the change of name on the official seal, many work units also used code names for the drugs developed. This is the reason for the numerous code names for the drugs that appear in the later part of the book.

(2) Research organisations

The main participating organisations were introduced in the research collaboration plans from 1967 to 1970.

1. Drug Synthesis and Screening Team
   The job description for this team was well defined. More than ten participating work units were involved in the chemical synthesis of the drugs : AMMS, Shanghai Pharmaceutical Factories No. 2 and No. 14, Shanghai Pharmaceuticals Holding Co., Ltd., Shanghai Pharmaceutical Industry Branch Office, Shanghai Institute of Pharmaceutical Industry, Shenyang Pharmaceutical Industry Branch Office, Institute of Materia Medica of the CAS, National Institute of Parasitic Diseases of the CAMS, Institute of Materia Medica of the CAMS, and the Chongqing Pharmaceutical Industry Branch Office. Each work unit was responsible for at least two research tasks. Shanghai Institute of Pharmaceutical Industry handled the highest number of research tasks, with six tasks in total. More than nine work units were involved in the drug screening: SMMU, the Third Military Medical University, AMMS, Institute of Materia Medica of the CAS, Institute of Parasitic Diseases of the CAMS, Institute of Materia Medica of the CAMS, Jiangsu Institute of Parasitic Diseases, Shandong Institute of Parasitic Diseases and the Sichuan Academy of Chinese Medicine Sciences. These work units handled one to two tasks each, and drugs which exhibited strong anti-malarial properties would be passed to the first six work units for pharmacology, pathology and toxicology experiments to further screen the drugs before the embarking on clinical trials. The drugs selected came mainly from the AMMS, Shanghai Pharmaceutical Factories No. 2 and No. 14, Shanghai Pharmaceutical Industry Branch office, Shanghai Institute of Pharmaceutical Industry, Shenyang Pharmaceutical Industry Branch office, National Institute of Parasitic Disease of the CAMS, Institute of Materia Medica of the CAMS, Chongqing Pharmaceutical Industry Branch office, and Beijing and Shenyang regions.

2. Traditional Chinese Medicine and Acupuncture Team
   This team handled three research topics. Eight of the participating work units, namely the Sichuan Academy of Chinese Medicine Science, Third Military Medical University, Jiangsu Wuxi Institute of Parasitic Diseases, Jiangsu Institute of Traditional Chinese Medicine, Institute of Materia Medica of the CAS, AMMS,

Shanghai Institute of Pharmaceutical Industry, and the Institute of Materia Medica of the CAMS were in charge of the research on *Dichroa febrifuga* and other Chinese herbs known to be effective against malaria. The medicines in common use that showed efficacy against malaria were investigated by the main participating work units as well as other relevant work units specially appointed by the science and technology commissions and health departments of the provinces of Yunnan, Guangxi, Guangdong and Hainan. Some external participating work units in the Yunnan region were the Chongqing Academy of Chinese Materia Medica, Shanghai Institute of Materia Medica, CAS, Shanghai Institute of Pharmaceutical Industry and AMMS. In the Guangxi region, there were the Nanjing Institute of Botany of the CAS, Shanghai Institute of Materia Medica of the CAS, Shanghai Institute of Pharmaceutical Industry and the AMMS. In the Guangdong and Hainan regions, there were the Beijing Institute of Materia Medica of the CAMS, Shanghai Institute of Pharmaceutical Industry and the AMMS. The third topic, acupuncture for malaria prevention and treatment, was mainly researched by the Shanghai Research Institute of Acupuncture, National Institute of Parasitic Diseases of the CAMS, acupuncture teaching and research groups of the Guangzhou University of Chinese Medicine and Nanjing University of Chinese Medicine, and the Northern Jiangsu People's Hospital.

3. Mosquito Repellants Team
   There were two research topics for this team. One of the topics was to investigate and improve the efficacy of mosquito repellants through field research. This was carried out mainly in Kunming and Guangxi. The investigation at the Kunming site was handled by the Military Medical Institute of the Logistics Department of the Kunming Military District. Other participating work units were the Third Military Medical University, the Xinan Branch Office of the CNPIC, logistics and health departments of the Nanjing Military District and AMMS. AMMS was responsible for the Guangxi site with participating personnel from the pharmaceutical department of the Shanghai Institute of Pharmaceutical Industry. The other research topic was topical and oral mosquito repel-

lants. This was mainly carried out by the Institute of Materia Medica of the CAMS, Kunming Institute of Botany of the CAS, South China Institute of Botany of the CAS, Institute of Botany of Jiangsu Province of the CAS, Kunming Institute of Chinese Medicine, Sichuan Institute of Chinese Medicine, Guangdong Institute of Chinese Medicine, Shanghai Institute of Pharmaceutical Industry, SMMU, Institute of Zoology of the CAS, Nanjing Pharmaceutical Factory, Chongqing Xinan Pharmaceutical Factory, Shanghai Pharmaceutical Industry Branch Office, Tianjin Pesticide Experimental Plant, Institute of Plant Physiology and Ecology of the Shanghai Institutes for Biological Sciences of the CAS, Third Military Medical University, AMMS, Shanghai Institute of Occupational Health, National Institutes for Food and Drug Control and the Shanghai Plastics Factory.

4. Drug Manufacturing and Packaging Team
   This team handled three research topics. Research into the production of tablets was mainly handled by the Shanghai Institute of Pharmaceutical Industry and Shanghai Pharmaceutical Factory No. 11, and they collaborated with Shanghai Zhongzhou Pharmaceutical Factory, Shanghai Institute of Parasitic Diseases, and SMMU. The research into injectable forms was mainly done by Shanghai Pharmaceutical Factory No. 10, AMMS and Shanghai Institute of Pharmaceutical Industry, and they collaborated with the Shanghai Institute of Parasitic Diseases, Shanghai Zhongzhou Pharmaceutical Factory and SMMU. Research into Chinese medicine formulations was carried out by the Shanghai Institute of Pharmaceutical Industry and Chongqing Academy of Chinese Materia Medica. They collaborated with the Shanghai Institute of Materia Medica, CAMS, Shanghai Institute of Parasitic Diseases, SMMU and Third Military Medical University.

5. Field Prevention and Treatment Team
   This team was divided into the Hainan, Kunming and Nanjing sites. They were responsible for the field testing of all the drugs. The main work units responsible were the Health and Epidemic Prevention Institute of the Logistics Department of the Guangzhou Military District, National Institute of Parasitic Diseases of the

CAMS, Hainan Military Logistics Health Office, Hainan Institute of Parasitic Diseases, Military Medical Institute of the Logistics Department of the Kunming Military District, Yunnan Institute of Parasitic Diseases, Health Office of the Logistics Department of the Nanjing Military District, Jiangsu Institute of Parasitic Diseases, Shanghai Research Institute of Acupuncture, AMMS, Jiangsu Institute of Traditional Chinese Medicine, Sichuan Academy of Chinese Medicine Sciences and Third Military Medical University.

In the plan drafted in 1967, there were more than fifty participating work units, and many units handled multiple research tasks at the same time. There were some work units not mentioned in the plan which still participated in some part of the research work. For instance, between 1967 to 1969, the work units present at the site of the National 523 Specialist Team field work included the Fourth Military Medical University, Institute of Health and Epidemic Prevention of the Guangzhou Military District, Hainan Military District Hospital No.187, Guangdong Institute of Parasitic Diseases, Epidemic Prevention Squad of the Hainan Military District, Hainan General Hospital, Guangzhou Zhongshan School of Medicine, Guang'anmen Hospital of the ATCM, Peking Anti-Imperialist Hospital, Institute of Military Medical Sciences of the Nanjing Military District, and Nanjing Military District Hospital No. 81. Subsequently, more units joined in as required. The participation of the ATCM only began in January 1969.

By the beginning of 1971, a total of more than seventy research organisations were participating in Project 523.

Mid-term change of leadership and participating organisations (1971–1981)

Most research work had ground to a halt at the height of the Cultural Revolution in 1967. Project 523, however, had been permitted to continue as it had been authorised by the highest leader. During that period, the operations of the national ministries and commissions, local government and research units underwent many upheavals with major changes in leadership and structure. Difficulties arose because many of the new leaders did not fully understand the importance of Project 523. After the expiry of the three-year anti-malaria research plan drafted in 1967, the Military Control Commission of the MOH, MCI, CAS and GLD submitted the *Anti-Malaria Research Report and Request* to the State Council and Central Military Commission on 16 March 1971. The report suggested making adjustments to the Leadership Group, appointing the MOH to be the team leader and the GLD to be the assistant team leader, while retaining the office at the AMMS. On 15 April 1971, the State Council and the Central Military Commission passed the (71) State Document No. 29 to approve the request. On 22 May 1971, the Anti-Malaria Research Meeting convened in Guangzhou. During the forum, the six original organisations of the Project 523 Leadership Group which consisted of STC (leader), GLD (assistant leader), COSTIND, MOH, MCI and CAS were changed to MOH (leader), Department of Health of the GLD (assistant leader), MCI and CAS. The four organisations were to take the lead while the office remained at the AMMS. In addition, the meeting drafted a five-year plan for national anti-malaria research from 1971 to 1975. The corresponding research plan and manpower needs were adjusted accordingly.

The writer saw a new stamp with "National Anti-Malaria Research Leadership Group" on a document that was released not long after the meeting. However, there was no official explanation given, so he consulted the former 523 Office staff, Shi Linrong.

Li: I saw that the seal of the Beijing Anti-Malaria Research Leadership Group had been amended to "Beijing 523 Leadership Group" instead of "Anti-Malaria Leadership Group Office" in June 1968. The Regional Anti-Malaria Research Leadership Group (codenamed Regional 523 Leadership Group) had their seal amended too. From the beginning of 1971, numerous seals with "National Anti-Malaria Research Leadership Group" emerged, and the use of the "Beijing 523 Leadership Group" seal decreased. As far as I could see, the Beijing 523 Leadership Group of the early days had been doing the same work as the National Anti-Malaria Research Leadership Group. The 1971 Guangzhou meeting reinforced the Leadership Group's role and the group was renamed the National Anti-Malaria Research Leadership Group. Just as in the other regions, there was a Regional 523 Office in Beijing.

Shi: The seal of the Anti-Malaria Drug Research Leadership Group Office was used after the Project 523 meeting in 1967. However, because Project 523 did not involve only drug research and to preserve secrecy in the chaotic situation in the country then, the seal was changed to "Beijing (National) 523 Leadership Group Office" after the 1968 meeting in Hangzhou. In order to differentiate each region from Beijing, "regional" was added to the names of the offices of each region. After 1971, some organisations and work units did not fully understand the function of Project 523. Also, as malaria became rampant in the country, some results from Project 523 research were widely used. As five provinces of the central plains (Shandong, Henan, Jiangsu, Anhui, Hubei) began clinical trials of a few drugs in Hubei, Henan and elsewhere, more seals of the National Anti-Malaria Research Leadership Group and Office were made. However, the Project 523 seal continued to be used, and two corresponding sets of stationery and envelopes were even printed. There was always a Leadership Group in the Beijing region, just that the office was directly governed by the National Office.

It can be observed that the Project 523 Leadership Office was constantly changing its policies to adapt to the changing situation in the country. After the 1971 meeting, other than making changes to the original research units, new work units were also invited to participate in the research to meet the new research requirements. These units were the National Vaccine and Serum Institute, Beijing Medical University and Beijing Institute of Pharmaceutical Industry. After the SMMU troops moved to Xi'an, Xi'an Pharmaceutical Factory also began collaborating in the research. Subsequently, the main participating work units of Project 523 collaborated with additional work units as the need arose.

According to the records, the Leadership Group comprising the four organisations was in charge until 1978, after which the State Administration of Medicine took over. On 4 September 1979, the State Administration of Medicine document (79) Drug No. 387 proposed that the drug synthesis research be handed over to the AMMS. From 1980 onwards, MCI was no longer in charge of the military medical research. Additionally, the proposal mentioned that Project 523 was carrying out a decreasing number of tasks and acknowledged that the project was a collaborative effort between both military and civilian organisations. It stated that, from 1980, all civilian medical research plans would no longer involve the military medical organisations. Subsequently, the Leadership Group was taken over by the MOH, STC, State Administration of Medicine and GLD, and the MCI and the CAS were no longer part of the Leadership Group.

At the end of March 1981 when the National Anti-Malaria Leadership Group was disbanded, the MOH, STC, State Administration of Medicine and GLD jointly issued awards to the participating work units and individuals within the Leadership Groups. The criteria for the awards were:

1) to have organised and led Project 523 research and achieved notable results;
2) to have obtained significant Project 523 research results (research units, participating units and major collaborating units);
3) to have engaged in Project 523 research over a prolonged period and obtained notable results in science and technology, product identification and scientific techniques;
4) to have achieved outstanding results in manufacturing and promoting the results of Project 523.

Note: If the Project 523 research task was completed with outstanding results, the award was presented to the work unit. If significant results were obtained, the award was presented to the individual research group and research laboratory.

A total of 134 work units (collective) received awards. 17 units were in the science and technology sector, 55 units were in the medical and health sector, 27 units were in the pharmaceutical and chemical sector, 26 units were in the military sector, and 9 units were in the light manufacturing, higher education and other sectors. A total of 85 individuals received awards; they came from seven regions: Beijing, Guangdong, Guangxi, Nanjing, Shanghai, Sichuan, and Yunnan.

The conclusion of Project 523

On 25 August 1980, the four leading institutions, the MOH, STC, State Administration of Medicine and GLD jointly requested that the State Council and Central Military Commission dissolve the National Anti-Malaria Drug Research Leadership Group and the national collaboration between organisations. This covered all personnel from the provincial, city and state commissions involved in anti-malaria research. The document was signed by Huang Shuze, Zhao Dongwan, Huang Kaiyun and He Biao, and reviewed by Qian Xinzhong. It was then approved by Deputy Prime Minister Chen Muhua, and read and agreed by Wan Li and Fang Yi. The document stated that the forum for the Anti-Malaria Drug Research Leadership Group and the personnel in charge of the regional offices, originally scheduled for the fourth quarter of 1980, would convene in 1981 instead. The forum was held from 3 to 6 March 1981 during which some changes were made to the list of Project 523 collaborating organisations. However, anti-malaria research would still remain a major focus of pharmaceutical and health research. It was also included in the regular research plans of relevant departments from the states, provinces, cities, autonomous regions and the military. In view of the changes to the list of Project 523 collaborating organisations, the MOH set up an Anti-Malaria Committee within the Medical Sciences Committee. The military also decided that the GLD would organise an Anti-Malaria Team within the Specialist Epidemics Team in May 1981. On 11 May 1981, representing the National Anti-Malaria Leadership Group, the four leading institutions jointly issued a final notice — the *Summary of the Anti-Malaria Drug Research*. Other than summarising the discussion at the forum, the notice also outlined the plan for closing down Project 523. It stated that the regional leadership groups would oversee the transfer of the Project 523 Office documents, technical files, funds and supplies. The leadership groups and work units involved would also need to make proper arrangements for the full-time Project 523 research and management personnel who had been away from their original work units for a long period of time. However, according to the work unit archives and recollections of some research personnel, many technical files were mishandled and lost.

By March 1981, the military and civilian collaboration under Project 523 had been mostly completed, but work related to Project 523 continued. The regional anti-malaria work was handed over to the Anti-Malaria Committee, and a corresponding Anti-Malaria Team was set up within the military. However, there were some issues that were not handled well. For example, there was no documentation about the award for discovering artemisinin and how it was named, and these issues are still the subject of dispute today. Subsequently, in order to adapt to new international circumstances, the cooperation between China, the World Health Organization (WHO) and other organisations was strengthened, and there was further collaboration on projects such as the research and development of artemisinin and its derivatives. At the suggestion of the WHO, on 20 March 1982, the MOH and State Administration of Medicine jointly set up the National Steering Committee for Development of Artemisinin and its Derivatives. Unfortunately, some of the internal reports published by the Anti-Malaria Committee did not contain much relevant information on artemisinin research. Of the more than seven hundred documents and abstracts published in two compilations of documents published in 1985 and 1990, only about twenty articles were related to artemisinin and its derivatives. The articles mostly reported the efficacy of artemisinin when administered together with other medications. This might be because the Steering Committee was responsible for developing artemisinin while the Anti-Malaria Committee was not really involved.

Note: The main materials used for this book originated from the period before 2011.

Tu Youyou received the Lasker Award in 2011.

A young Tu Youyou

Tu Youyou in her teens

COMPARISON OF THE STRUCTURE OF QINGHAOSU WITH OTHER ANTI-MALARIA DRUGS

Compounds	Structural formular
Quinine	
Dichroine	
Atebrin	
Chloroguanidine	
Pyrimethamine	
8-Aminoquinoline	
Chloroquine	
Primaquine	
Qinghaosu	

青蒿素

双氢青蒿素

The molecular structure of artemisinin and dihydroartemisinin

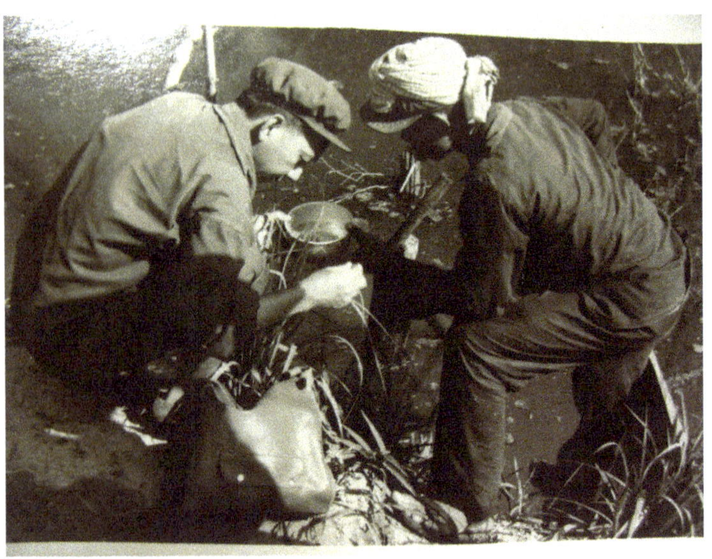

Searching for mosquitoes in Cambodia in 1976

Chapter 2
The Discovery of Artemisinin

This honour is not mine alone, but also belongs to my team and all my comrades in China.

This is a successful example of collective research in traditional Chinese medicine, and is an honour for the globalisation of Chinese science and Chinese medicine.

— Tu Youyou

Rediscovering the Anti-Malarial Properties of *Artemisia Annua*

Project 523, or the Anti-Malaria Research Work Plan, contained five research topics. The second topic in the plan was to research traditional Chinese medicine and acupuncture remedies that were reputed to show efficacy against malaria. After searching the literature, *Artemisia annua* (*A. annua*) was fifth on the list of medicines to be researched. However, no further record was found regarding the screening of *A. annua*. A number of researchers remembered performing the preliminary screenings, but since numerous herbs were screened, and many of them presented similar anti-pyretic effect against malaria, unless the herb presented extremely outstanding possibilities, it would most likely be overlooked.

The Academy of Traditional Chinese Medicine (ATCM) began its involvement in Project 523 in January 1969. Subsequently, in April 1969, a book named *A Collection of Secret Anti-Malaria Prescriptions* (Figure 6)

疟疾单秘验方集

内部参考资料

一九六九年

Figure 6: A Collection of Secret Anti-Malaria Prescriptions

was compiled under the name of ATCM Revolutionary Committee working group. A paragraph on page 15 (Figure 7) contains information on *A. annua*. It reads:

> *"Prescription: 15 g to 250 g of A. annua;*
> *Method: crush to extract the juice and drink, or drink the water decoction, or grind into fine powder, mix with hot water and drink;*
> *Source of herb: Fujian, Guizhou, Yunnan, Guangxi, Hunan, Jiangxi."*

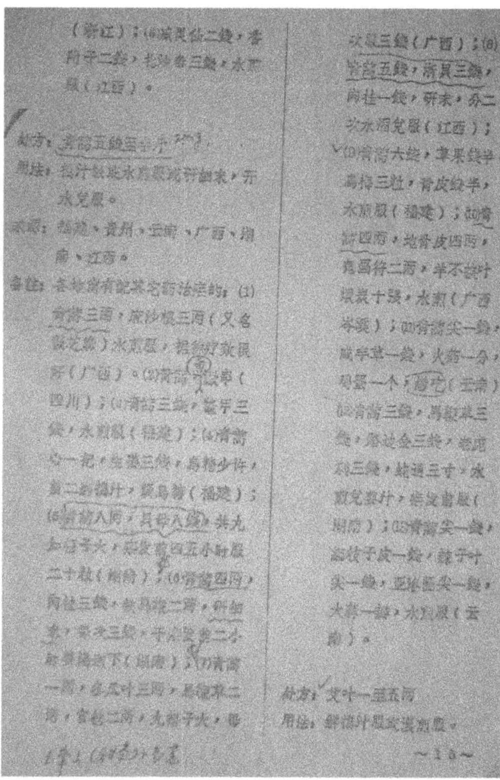

Figure 7: *A page on A. annua in A Collection of Secret Anti-Malaria Prescriptions*

The book also described how *A. annua* was used in different regions, and how it could be used together with other herbs to treat malaria. A total of 13 prescriptions involving *A. annua* were recorded in the book.

However, in 1969, instead of researching on the anti-malarial properties of *A. annua*, the ATCM focused their research on Drug No. 52. In 1970, researcher Xu Yagang from the ATCM revisited the Chinese medical literature to make a further selection of medicines from the literature. One of the more important reference books used was *Monograph on Malaria* published by the Shanghai Literature Institute of Traditional Chinese Medicine (Figures 8–9). He made a list of 808 prescriptions from the book, of which 519 prescriptions from *Monograph on Malaria* as well as 55 prescriptions from the *Integrated Medical Department Records* did not contain *Dichroa febrifuga*. Hence, there were a total of 574 prescriptions to analyse. Subsequently, the individual medicines were listed (Figure 10) for analysis and

Figure 8: Cover of Monograph on Malaria

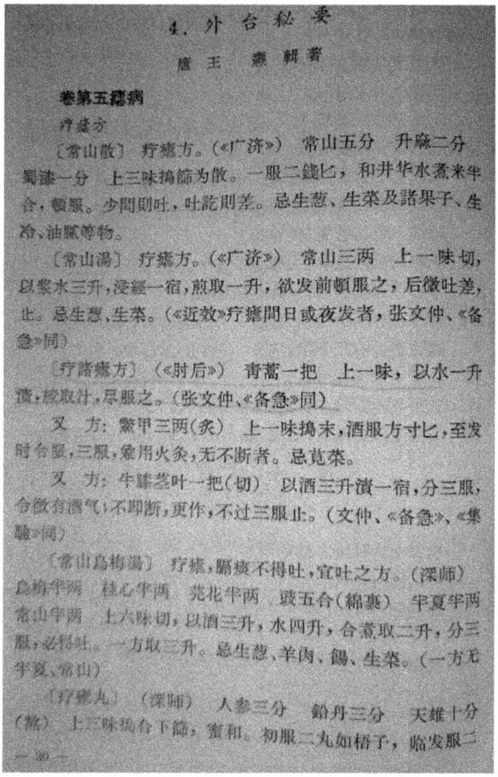

4. 外台秘要

唐王 焘 輯書

卷第五瘧病

疗瘧方

〔常山散〕疗瘧方。(《广济》) 常山五分 升麻二分 蜀漆一分 上三味捣篩为散。一服二錢匕，和井华水煮米半合，顿服。少間則吐，吐訖則差。忌生葱、生菜及諸果子、生冷、油膩等物。

〔常山湯〕疗瘧方。(《广济》) 常山三两 上一味切，以浆水三升，浸經一宿，煎取一升，欲发前頓服之，后微吐差，止。忌生葱、生菜。(《近效》疗瘧間日或夜发者，张文仲、《备急》同)

〔疗諸瘧方〕(《肘后》) 青蒿一把 上一味，以水一升漬，絞取汁，尽服之。(张文仲、《备急》同)

又方：鱉甲三两(炙) 上一味捣末，酒服方寸匕，至发时令服，三服，兼用火灸，无不断者。忌莧菜。

又方：牛膝茎叶一把(切) 以酒三升漬一宿，分三服，令微有酒气，不即断，更作，不过三服止。(文仲、《备急》、《集験》同)

〔常山烏梅湯〕疗瘧，膈痰不得吐，宜吐之方。(深师) 烏梅半两 桂心半两 芫花半两 豉五合(綿裹) 半夏半两 常山半两 上六味切，以酒三升，水四升，合煮取二升，分三服，必得吐。一方取三升。忌生葱、羊肉、餳、生菜。(一方无半夏、常山)

〔疗瘧丸〕(深师) 人参三分 鉛丹三分 天雄十分(炮) 上三味捣合下篩，蜜和。初服二丸如梧子，临发服二

— 30 —

Figure 9: A page from Monograph on Malaria

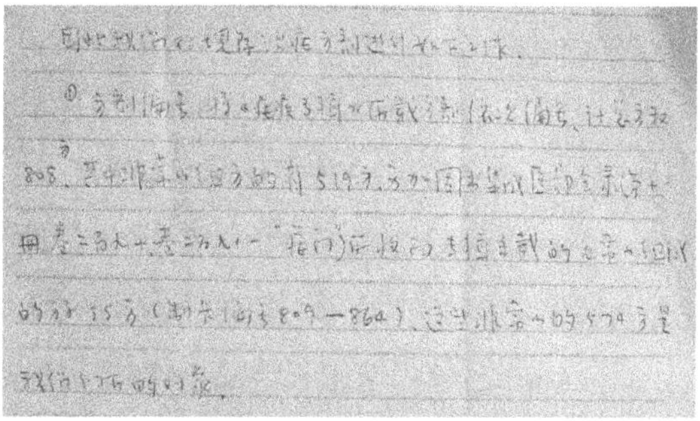

Figure 10: Part of the information gathered by Xu Yagang

sorting. Prescriptions found in the *Taiping Benevolent Dispensary Bureau Prescriptions* were in one list, and those not included were in a second list. Some medicines were used on their own, while others were used in combination with other medicines. After removing the medicines which were duplicated in the two lists, the results were tabulated in a third list. The medicines selected as the focus for research were: *Aconitum*, *Fructus mume*, *Trionyx sinensis* shell and *A. annua*. These medicines were chosen because there were records of their use either on their own or in combination with other medicines, and there was some basis to justify their being tested on animals. These medicines were sent to the AMMS for testing on malaria-infected mice. Approximately one hundred prescriptions were screened and the extracts of *A. annua* showed a 60–80% success rate against malaria parasites. However, the results were inconsistent.

According to the researchers, the three solvents used for extractions were water, ester and alcohol. The alcohol extracts of *A. annua* showed efficacy against malaria parasites, but the results were not consistent. However, after repeated experiments, the alcohol extracts showed a success rate as high as 90%. In addition, another medicine, realgar, an arsenic sulfide mineral, showed a success rate as high as 99%. Initially, Xu Yagang placed his main emphasis on realgar, but researcher Gu Guoming from AMMS raised an objection, pointing out that realgar, when heated to a certain temperature, could be oxidised to a highly toxic vapour, arsenic trioxide. This made realgar unsuitable for common use, especially in the military, and the central government would not be likely to approve its use. Thus, the research on realgar was halted. Following this, Xu Yagang shifted his attention to *A. annua* as it exhibited the second highest success rate. He presented his results to Tu Youyou, his research leader. Some time later, Tu Youyou instructed that the following medicines be screened (Figure 11):

Mineral-based medicines: minium, realgar, sulphur, cinnabar and ferrous sulphate.

Animal-based medicines: *armadillidium vulgare*, *pheretima*, snake slough, pangolin, and eggshell membrane.

Plant-based medicines: *Cortex lycii*, *Euphorbia kansui*, *Hemerocallis citrina*, *Brucea javanica*, *A. annua* and *Verbena officinalis*.

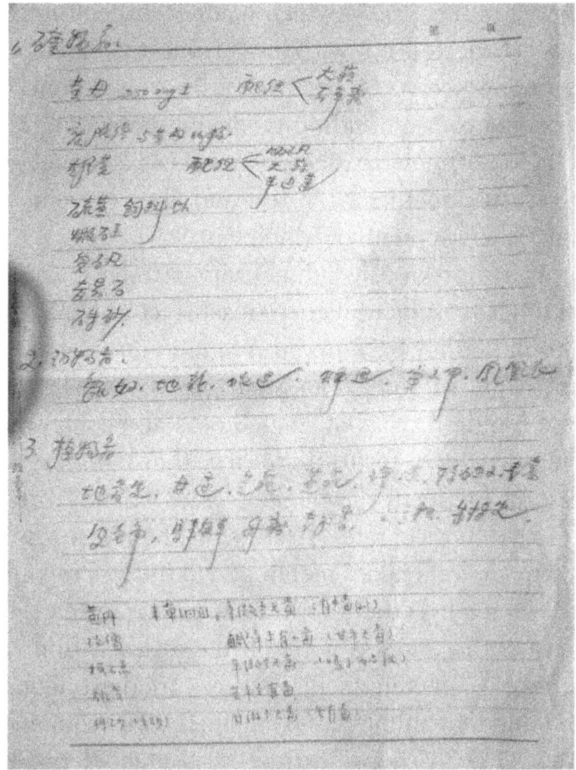

Figure 11: A list of medicines for screening compiled by Tu Youyou

According to information gathered by Xu Yagang, (Figure 10), minium and other mineral-based medicines in combination with other medicines were the main research focus. At the beginning of 1971 (the New Year was spent on the train), Xu Yagang was transferred out of the Project 523 team and assigned to the Bronchitis office in the Northeast region with immediate effect.

On 22 May 1971, the National Anti-Malaria Research Forum convened in Guangzhou. During this forum, the six original leading organisations, STC (leader), GLD (assistant leader), COSTIND, MOH, MCI, and CAS were changed to the MOH (leader), Health Department of the GLD (assistant leader), MCI and CAS. The forum also drew up a five-year plan (1971–1975) for Anti-Malaria Drug Research. Due to a shortage of manpower in the ATCM which made it difficult to carry out the research, they requested to withdraw from this area of research. However, this was rejected by the MOH. Tu Youyou once mentioned that after screening over a hundred medicines, including *Piper nigrum* extracts which did not show satisfactory results against malaria parasites, they had to seek new medicines to test and also revisit medicines that had previously shown relatively high success rates. *A. annua* was re-selected as it had previously been observed to have a success rate of 68%. As further test results remained unsatisfactory with a success rate of only 12% to 40%, they gave up on *A. annua* once again. During the second half of 1971, Tu Youyou and her team from ATCM once again selected over a hundred types of single herbs and compound prescriptions, including *A. annua,* for testing. First, the *A. annua* water decoction was deemed ineffective, while the 95% ethanol extraction only yielded a 30–40% success rate. Tu Youyou then was inspired by the instruction "wring out the juice" mentioned in the folk remedies. Hence, she suggested using ethyl ether to extract *A. annua* instead. This method greatly increased the success rate of the extract from 30–40% to 95–100% in malaria-infected mice. They also tried extraction with ethanol, but they found that although the ethanol extract contained the same compounds as the ethyl ether extract, it contained two-thirds more impurities than the latter, thus affecting its efficacy. After further purification to remove the ineffective acidic components present in the extract, an effective neutral compound was obtained. By the end of December, the tests on malaria-infected monkeys achieved similar results as the tests on mice. Tu Youyou ascribed her inspiration to Ge Hong's *Handbook of Prescriptions for Emergencies*, which mentioned the experience of squeezing out juice from *A. annua*. The instruction: "A handful of *qinghao* immersed in a litre of water, wring out the juice and drink it all" caused Tu Youyou to consider whether high temperatures or enzymatic destruction might be the problem. After switching to extraction with ethyl ether, with a lower boiling point than ethanol, and separating the acidic from the neutral components, on 4 October 1971, after numerous experiments, extract No. 191 (No. 91 in some documents) was extracted from *A. annua*. This neutral extract sample exhibited a 100% success rate against the malaria parasites in infected mice.

The Extraction and Naming of the Effective Monomer of *A. Annua*

After the ATCM first achieved conclusive results of the efficacy of *A. annua* extract against malaria, the Shandong Academy of Chinese Medicine and the Yunnan Institute of Materia Medica began conducting similar experiments. The research is summarised below.

Academy of Traditional Chinese Medicine (ATCM)

From the latter half of 1971, although the *A. annua* ethyl ether extracts obtained exhibited a 100% success rate against the malaria parasites in malaria-infected mice, the researchers realised that it had rather high toxicity and low safety levels. There was thus a need to decrease the toxicity of the extract while finding ways to increase its potency. After further separation and extraction experiments, they discovered that the ineffective and more toxic portion was the acidic component which could be removed by sodium hydroxide. The remaining neutral extract exhibited a significant lower toxicity and higher potency.

The National 523 Office convened meetings in Nanjing in March 1972 for the Drug Synthesis and Screening and Traditional Chinese Medicine and Acupuncture Teams. During the Traditional Chinese Medicine and Acupuncture Team meeting, Tu Youyou reported the results of recent research on *A. annua* which showed a 100% success rate in malaria-infected mice. According to the minutes of the meeting, the research into many other medicines was also presented:

After exhibiting promising results against malaria in clinical trials, the chemical structure of *Artabotrys hexapetalus* was being analysed. Preliminary separation of the effective chemical monomer of *Agrimonia pilosa Ledeb.* had begun and work to determine its chemical structure would be undertaken if it showed effective results in clinical trials. The effective elements of the following herbs were extracted: *A. annua*, *Ailanthus altissima*, *Strigose hydrangea*, *Hydrangea macrophylla*, *Nandina domestica*, *Aristolochia yunnanensis Franch.* and *Euphorbia helioscopia*. Most of these herbs were undergoing laboratory experiments and clinical research to improve their potency. The meeting suggested the following actions: to quickly determine the chemical structure of *Artabotrys hexapetalus* and to research its synthesis; to confirm the clinical efficacy of the effective monomer of *Agrimonia pilosa Ledeb.* and determine its chemical structure; to study the clinical effectiveness of A. *annua*, *Ailanthus altissima* and other selected herbs, and concurrently work on the extraction of the effective compounds or monomers.

This was the first mention of the anti-malarial properties of A. *annua* extracts at a Project 523 specialist team meeting. By the end of 1972, ATCM had separated three types of crystals from the neutral compound of the A. *annua* ethyl ether extract. The crystals separated were arteannuin-A, artemisinin and arteannuin-B. The first two crystals were white needle-like crystals while arteannuin-B was a cubic crystal. The crystal which exhibited anti-malarial properties was artemisinin. The timeline of the extraction of artemisinin and the naming of the compound were disputed: Tu Youyou recalled that the separation of the crystals was done on 8 November 1972, and the effective monomer was artemisinin. According to the book *A Detailed Chronological Record of Project 523 and the Discovery and Development of Qinghaosu (Artemisinin)* written by Zhang Jianfang (published in 2006 by Yangcheng Evening News Publishing House), the ATCM separated multiple monomers from A. *annua* and one of the effective anti-malarial monomers was named "artemisinin II". Other information from the ATCM showed that an effective monomer "artemisinin II" was refined from the neutral A. *annua* ethyl ether extract. After the purification of artemisinin II, silica gel column chromatography was used to produce a cubic crystal that was tentatively named "artemisinin III". There were other records which mentioned that the effective monomer obtained by the ATCM then was "artemisinin II". The academy carried out clinical trials using this monomer and also began analysis of its chemical structure at the Shanghai Institute of Organic Chemistry in February 1974. Right up to 1975, throughout the clinical trials, the name "artemisinin II" was used instead of "artemisinin".

To find out more about the naming of artemisinin, the author interviewed some of the researchers from the ATCM. One of them recalled:

After joining Tu Youyou's research team, Ni Muyun designed a procedure to process the extracts before column separation. First, the neutral *A. annua* ethyl extracts would be evenly mixed with polyamide before percolation using 47% ethanol. After concentrating the filtrate, extraction with ethyl ether would be performed. While the efficacy of the concentrated ethyl ether extract against malaria improved further, no solid substance was obtained when they tried to separate the extracts using alumina column chromatography. Zhong Yurong then suggested that silica gel column chromatography might be a more suitable option to separate the neutral compound. Employing gradient elution using petroleum ether and ethyl acetate petroleum ether, a large amount of cubic crystals was produced. The crystals were named "crystal I"; next, a small quantity of needle-like crystals were produced, numbered "crystal II"; the last group of crystals produced were another type of needle-like crystals, numbered "crystal III". After testing the crystals on malaria-infected mice, only "crystal II" was found to be effective against malaria parasites. All subsequent reports made to the National 523 Office used the name "artemisinin II" instead of "crystal II".

After looking at all the accounts from different sources, the author is inclined to the view that the name "crystal II" was changed to "artemisinin II". In addition, many others recalled using the name "artemisinin II" in the early stages. Furthermore, "artemisinin II" was used in the diaries kept by some of the researchers.

Shandong Institute of Parasitic Diseases and Shandong Academy of Chinese Medicine

After attending the Project 523 Traditional Chinese Medicine Specialist Team meeting held in Nanjing in March 1972 and hearing about the research findings of the ATCM, Shandong Institute of Parasitic Diseases used the ethyl and alcohol extracts of *A. annua* produced in Shandong province to carry out malaria treatment experiments on animals. They achieved excellent results and sent a written report to the National 523 Office on 21 October 1972. The experimental results indicated that the extracts from *A. annua* L. achieved the same anti-malarial effect in malaria-infected mice as described in the experimental report presented by the ATCM. Subsequently, Shandong Institute of Parasitic Diseases collaborated with Shandong Academy of Chinese Medicine. From October 1973 onwards, they started work on separating the effective monomer but as there was a shortage of researchers, only two researchers worked on this project. In November 1973, Shandong Academy of Chinese Medicine succeeded in extracting seven types of crystals from the *A. annua* L. collected from Tai'an, Shandong province. The No. 5 crystal extract was found to be an effective anti-malarial crystal and was named "*huanghuahaosu*".

Yunnan Institute of Materia Medica

At the end of 1972, director Fu Shuliang of Yunnan Regional 523 Office attended the annual Regional 523 Office management meeting in Beijing. After learning about the ATCM's *A. annua* research results, he held a meeting to disseminate this information to the researchers in the Yunnan Institute of Materia Medica. He instructed them to make use the favourable conditions in the region to produce and screen members of the *Asteraceae Artemisia* species. During spring 1973, Luo Zeyuan, a researcher from the Yunnan Institute of Materia Medica, discovered a type of pungent, foot-tall plant in the campus of Yunnan University. She plucked and dried the plants and began the extraction process. She did not recognise the plant until Liu Yuanfang, a botanist, told her that the plant she found was *Artemisia absinthium*. It was found that the ethyl ether extracts of *Artemisia absinthium* exhbited anti-malarial properties with reproducible results. Subsequently, screening and extraction were carried out concurrently and, by April 1973, the effective anti-malarial monomer was isolated and given the temporary name "*Artemisia absinthium* crystal III". The name was later changed to "*huanghaosu*". Not long after this, Luo Kaijun of the Institute sent an *Artemisia absinthium* plant specimen to a botany professor Wu Zhengyi from the Kunming Institute of Botany, CAS for evaluation. He confirmed that the scientific name of plant was *Artermisia annua* L. f. *macrocephala* Pamp. Subsequently, a type of *A. annua* plant was purchased from a Chinese medicinal herb company in Chongqing, Sichuan. The plant was *Artemisia annua* L., and yielded an even larger quantity of *huanghaosu*.

The description of the research process of Shandong and Yunnan in the book *A Detailed Chronological Record of Project 523 and the Discovery and Development of Qinghaosu* (*Artemisinin*) is similar to that presented at the 28 February 1974 meeting of the ATCM.

In conclusion, it can be confirmed that after Tu Youyou revealed the anti-malarial properties of *A. annua* ethyl ether extract, the Shandong Institute of Parasitic Diseases and Shandong Academy of Chinese Medicine together, and the Yunnan Institute of Materia Medica independently, began work on extracting an anti-malarial monomer from *A. annua*. Both institutes were able to obtain an effective anti-malarial monomer and named it *huanghuahaosu* (Shandong) and *huanghaosu* (Yunnan) respectively. Since the chemical structure had not been ascertained, and both the effective anti-malarial extracts came from the plant *A. annua* L. (or its variant), both names were associated with *A. annua* L. The extract that the ATCM named "artemisinin II" had been verified in February 1974, the two compounds known as *huanghuahaosu* and *huanghaosu* were regarded as the same compound, now known as artemisinin. Due to the difference in plant characteristics between *Artemisia annua* L. and *A. annua* and the research emphasis of the various institutes, it was quite natural that the effective monomers extracted were given different names. Many people have already researched this issue, and this book will not elaborate further. According to the author's research, the name given to the effective monomer extracted by these three institutes was not unified for a long time. Although the compound was finally unified as "artemisinin", questions are still raised when people recall the differences in the results of the clinical trials conducted by the three institutes and the differences in the original names. This will be further elaborated on in Chapter 4.

The Structure of Artemisinin

By the end of 1972, the ATCM had separated different crystals from *A. annua*. In 1973, they started to analyse the structure of artemisinin II. Tu Youyou's research team determined that artemisinin was a white needle-like crystal with a melting point of 156–157°C and optical rotation $[\alpha]_D^{17} = +66.3$ ($c = 1.64$ in chloroform). After doing the chemical reaction test, it was confirmed that artemisinin did not contain any nitrogen or double bonds and its elemental analysis was C63.72%, H7.86%. The team also carried out four large spectrum analyses using their own instruments as well as instruments from other units to determine the molecular formula of artemisinin which was confirmed to be $C_{15}H_{22}O_5$ with a molecular weight of 282. Subsequently, under the supervision of the late Professor Lin Qishou from the Beijing Medical University, they deduced that artemisinin II might be a type of sesquiterpene lactone, which was a new type of structure for an anti-malarial drug. As the terpene structure was a relatively new type of structure and the chemical research facilities of the ATCM were insufficient, it was difficult for them to carry out the chemical analysis. At the time, only a handful of scientists were researching chemical compounds. From published articles, they discovered that Professor Liu Zhujin from the Shanghai Institute of Organic Chemistry (SIOC) specialised in terpene structures and had much relevant experience in this area of research. Hence, the ATCM contacted Professor Liu to suggest a collaboration in performing the chemical analysis and to determine the structure of artemisinin. Tu Youyou brought the relevant documents to the SIOC and was received by comrade Chen Yuqun. In January 1974, Chen Fuhan agreed that someone from the ATCM would be designated to work together with the SIOC.

According to information collected by the authors, at the forum for the Anti-Malaria Drug Research Leadership Group and personnel in charge held in Shanghai from 28 May to 7 June 1973, the leadership group gave clear instructions for working on the analysis of the chemical structure of the active anti-malarial ingredient in A. *annua*. Researchers were told that while working on improving the formulation of artemisinin and promoting its use, there was a need to strengthen the collaboration between the organisations and so as to determine the chemical structure and carry out chemical synthesis research by 1974. As the authors were not able to look at the ACTM's original records, it cannot be confirmed whether the ATCM began research into the chemical structure of artemisinin before or after the meeting.

In February 1974, the ATCM sent Ni Muyun over to the SIOC, bringing along some research materials and a quantity of artemisinin. Wu Zhaohua recalls:

Wu: At that time, Liu Zhujin had had begun his work on liquid crystals, so he passed the artemisinin work to Zhou Weishan. Zhou Weishan was in charge of Laboratory 1, and since he had his own on-going work, he then passed the main artemisinin research to Wu Zhaohua who was from his lab. Wu Zhaohua would report the experimental findings to Zhou Weishan who would then discuss the research with everyone during the lunch break or after work. At that time, room 101 of Laboratory 1 was located on the second level of Building 1. Wu Zhaohua worked in the big Laboratory 227 and Wu Yulin worked in Laboratory 219. Many people would pass through the big laboratory and Wu Yulin often went to the big laboratory to socialise, so both of them were very familiar with each other. As Wu Zhaohua was unfamiliar with the new Magnetic Resonance Imaging (MRI) technique used for identifying chemical compounds, he would show the results to Wu Yulin and ask for his guidance.

Ni Muyun started working with Wu Zhaohua upon arrival at the SIOC. In the beginning, they mainly repeated the experiments performed in Beijing and also did some research on chemical reactions and spectral data. From February 1974, the ATCM sent a succession of researchers to the SIOC to work on the analysis of the structure of artemisinin II: Ni Muyun (February 1974 to the beginning of 1975), ZhongYurong (a short period of two to three months), Fan Jufen and Liu Jingming. The researcher at SIOC would report the progress to Tu Youyou in Beijing who would then discuss the results with either Lin Qishou or Professor Liang Xiaotian before sending feedback about the research to Shanghai.

While the ATCM and SIOC were collaborating on the analysis of the chemical structure, TuYouyou also hosted a collaboration with the Institute of Biophysics of the CAS in 1974 in Beijing, making use of an advanced X-ray diffraction method to help determine the chemical structure of artemisinin.

In April 1975, when Li Ying of the Shanghai Institute of Pharmaceutical Industry attended the Project 523 Traditional Chinese Medicine Forum in Chengdu, he heard Yu Dequan from the ATCM report on the chemical structure of *yingzhaosu*-A (another anti-malaria monomer which contained the peroxide group). He was inspired by this presentation and told Wu Yulin about it on his return to Shanghai. Wu Yulin then speculated that artemisinin might be a peroxide compound too. Subsequently, through qualitative and quantitative analysis, it was confirmed that artemisinin was indeed a peroxide compound. Considering experiments performed in Yugoslavia where cadinane sesquiterpene and arteannuin-B structures were extracted from the same plant, Wu Yulin proposed that the peroxide group might be present in the lactone ring. This provided a useful reference for the Institute of Biophysics in working out the structure of artemisinin. The complete and detailed structure of artemisinin was inferred by the Institute of Biophysics from its chemical structure through the use of the single-crystal X-ray diffraction analysis at the end of 1975. In 1978, anomalous X-ray diffraction scattering analysis was used to further confirm the molecular structure of artemisinin.

According to relevant records, there was a period of time when the ATCM was unable to extract any artemisinin. However, the National 523 Office arranged for the Yunnan and Shandong Institutes to provide some crystals of higher purity for the SIOC to continue the work of determining the chemical structure of artemisinin.

Throughout the process of determining the structure of artemisinin, the relevant institutes communicated and exchanged much information. Much work was done by each organisation to determine the structure. As the research conditions in China back then were rather undeveloped, many participating work units did not have the necessary instruments and equipment needed for the research. Under the coordination and organisation of 523 Office, almost all the most modern instruments and equipment in China were mobilised to carry out this work. It was the effort of the whole Project 523 team and the collaborating work units that made it possible to successfully determine the molecular structure of artemisinin despite the backward conditions then.

Clinical Trials

Under the special circumstances then, the ultimate goal was to find a drug that could be clinically used against *Plasmodium falciparum* (*P. falciparum*) as soon as possible. Therefore, all research was directed to support clinical use. Only after being proven effective in clinical trials could a drug become a true medicine. Hence, each institute completed their own clinical trials of their A. *annua* (A. *annua* L.) extract.

The neutral component of *A. annua* ethyl ether extracts

The ATCM was the first institution to carry out clinical testing of the unrefined extracts of A. *annua*. In 1972, from 24 August to the beginning of October, researchers from the ATCM conducted a clinical trial in the low-endemic region of Changjiang, Hainan, using the neutral extracts (No. 91) of A. *annua* ethyl ether extracts on the local population and those from other areas. The clinical trial consisted of 11 cases of tertian malaria, 9 cases of *P. falciparum* malaria, and 1 case of mixed infection. Chloroquine was used to treat 3 cases of *P. falciparum*. The tertian malaria cases were regarded as the controls.

The clinical definitions used:

Full recovery: Symptoms were controlled within 72 hours, blood films showed negative for *Plasmodium* parasites, no recurrence of symptoms within 7 to 12 days after discharge (patients were usually hospitalised for 4–5 days) and blood films showed negative for *Plasmodium* parasites.

Effective: Symptoms were controlled within 72 hours, blood films showed negative or remained positive for *Plasmodium* parasites, symptoms and *Plasmodium* parasites on blood film reoccurred 7 to 12 days after discharge.

Ineffective: Symptoms were not controlled within 72 hours, blood films showed positive for *Plasmodium* parasites.

At that time, little was known about A. *annua*, and the clinical effects of A. *annua* were still in the very early stages of exploration.

The results of the clinical trials of the efficacy of the neutral A. *annua* ethyl ether extract at Changjiang, Hainan are shown in Table 1.

Table 1: Clinical trials of the efficacy of the neutral *A. annua* ethyl ether extract at Changjiang, Hainan, 1972

Malaria type	Dosage: 3g per dose	Total No. of cases	Have a history of malaria or from malaria-endemic region	Time taken for fever to subside	Average time for fever to subside (hrs)	Number of days before *Plasmodium* turns negative	Effectiveness			Relapse	Remarks
							Full recovery	Effective	Ineffective		
Tertian malaria	Bid, consecutive 3 days	1	1	16°	36° 20′	5		1		2	1 case was not reviewed
		2		46° 30′		5	1	1			
	Tid, consecutive 3 days	1	4	8° 20′	11° 23′	2	1			2	
		3		12° 25′		2.3	1	2			
	Qid, consecutive 3 days	3	1	16°	19° 6′	2	2	1		1	
		1		27° 36′		2	1				
P. falciparum malaria	Bid, consecutive 3 days	1	Local		39° 50′	5	1				Able to control *Plasmodium* but unable to kill it completely
		1	Low malaria endemic region						1		
	Tid, consecutive 3 days	1	Local		24°	4	1				Able to control *Plasmodium* but unable to kill it completely
		1	Non-local						1		
	Qid, consecutive 3 days	5	Non-local		35° 9′	1.75	1	4		4	

The first clinical trial in Changjiang, Hainan proved that drug No. 91 had therapeutic effects on both tertian malaria and *P. falciparum* malaria on the local and non-local population in a low-endemic region. This was especially clear for tertian malaria patients where the efficacy rate was 100%. Furthermore, the groups given higher dosages exhibited better results and a lower rate of relapse. One patient in the low-endemic region with *P. falciparum* malaria did not react to the first treatment regime. The second treatment regime was ineffective for 1 out of 6 cases among the non-local population. Ignoring problems with the dosage or whether the patients were immune, there were a total of two ineffective cases for *P. falciparum* malaria. However, for the one case of mixed infection, no relevant record was found. Could the mixed case be one of the 11 cases of tertian malaria or was another person present? Since the clinical trial took place some time ago, the participants were unable to recall clearly. At the beginning of the records it was stated that there were 21 cases, but at the end it read:

> "A total of 11 tertian malaria cases were tested with the three types of treatment plans and 100% efficacy was achieved. One of them was infected with both tertian and *P. falciparum* malaria, but the main symptoms shown were due to tertian malaria. Hence, it was grouped together with the tertian malaria cases."

From these words and the table, it seems that the total number of cases treated was 20 and not 21.

There is also a record that the neutral *A. annua* ethyl ether extract was used in the 302 Military Hospital of China to treat nine cases of tertian malaria with a 100% efficacy rate. Therefore, speaking of efficacy alone, the results of the 1972 clinical trials showed that the neutral *A. annua* ethyl ether extract was effective in treating malaria.

In 1973, the Shandong Academy of Chinese Medicine and Shandong Institute of Parasitic Diseases performed a preliminary clinical trial using the effective compound "*huang* No. 1" which they had extracted from *A. annua* L. to treat 30 cases of tertian malaria. They discovered that this extract was effective in killing the *Plasmodium* parasites but showed a high relapse rate. They concluded that this drug was a fast-working but short-acting drug.

A. annua (*A. annua* L.) crystal extracts

From September to October 1973, the ATCM carried out clinical trials in Changjiang, Hainan using the artemisinin they had extracted. Out of the eight cases of tertian malaria and *P. falciparum* malaria they observed, three cases were among the non-local population. The dosage per capsule was 3–3.5g, and the average time taken for the fever to subside was 30 hours. Reviews were carried out after three weeks: two achieved full recovery, and treatment was effective in one case (*Plasmodium* parasites resurfaced 13 days later). For the five *P. falciparum* cases among the non-local population, treatment was effective for one case (more than 70,000/mm^3 of *Plasmodium* parasites, dosage was 4.5g in tablet form, the fever subsided in 37 hours, presence of parasites in blood films was negative after 65 hours, *Plasmodium* parasites resurfaced six days later); due to premature heart contractions, two cases stopped the medication early (this was the first time for one of the cases, with 30,000/mm^3 *Plasmodium* parasites, a dosage of 3g, fever subsided in 32 hours, *Plasmodium* parasites and fever resurfaced one day after stopping the medication); treatment was ineffective in the other two cases.

The Shandong *A. annua* L. Research Collaboration Team carried out their clinical observations in May 1974 at the ZhuZhuang Brigade of the Guandong commune, Juye County, Shandong. *Huanghuahaosu* was used on 10 cases of tertian malaria. The drug was given in capsule form, each capsule containing 0.1g of the crystal. The cases were split into two groups, each comprising five people. One group consisted of three adults and two children between the age 10 and 12. Dosage used: 0.2g for adult, 0.1g, for children once per day, to be consumed over three consecutive days. The other group consisted of five adults, dosage was 0.4g once per day, to be consumed for three consecutive days. Table 2 shows the rate of control of the symptoms and eradication of the *Plasmodium* parasites in the blood for each treatment group:

The conclusion from this first clinical trial performed by the Shandong Province *A. annua* L. Research Collaboration Team was that *huanghuahaosu* was a fast-acting anti-malarial drug which could be used for emergency treatment. There were no adverse effects observed during the treatment process. However, it was not a complete cure as the relapse rate was fairly high. As it would not be easy to control the rate

Table 2: Efficacy against symptoms and the *Plasmodium* parasites for each treatment group in 1974

Group	Dosage given	No. of cases	Average time taken for *Plasmodium* parasites to disappear from the blood (h)	Resurface of symptoms (days)			Reappearance of *Plasmodium* parasites in the blood		
				15	16–20	30–60	Day 15	Day 30	Day 60
1	*Huanghuahaosu* 0.2g × 3d	5	48	0	2	0	2	0	0
2	*Huanghuahaosu* 0.4g × 3d	5	33.6	1	2	0	1	0	0
3	*Huanghuahaosu* 0.4g with malaria prevention drug No. 2 taken periodically	9	50.7	0	1	1	0	0	2

of relapse with *huanghuahaosu* alone, it would be necessary to use *huanghuahaosu* concurrently with other anti-malarial drugs. This conclusion was similar to results of the clinical trials performed using simpler forms of *huanghuahaosu*.

On 7 September 1974, the staff of the Yunnan clinical collaboration team brought *huanghaosu* extracts to the Yun County for clinical trials. At that time, the weather had already turned cold, and the number of malaria cases had also dwindled. Hence, only one case of tertian malaria was treated. On 6 October, Liu Pu from the ATCM joined the collaboration team as an observer. Later, after communication with the Gengma County Epidemic Prevention Station, they were told that there were *P. falciparum* cases there. So Lu Weidong, Wang Xuezhong and Liu Pu went together to Gengma County on 13 October to carry out a clinical trial. There, they met a team from Guangdong, led by Li Guoqiao, which was setting up a treatment centre for cerebral malaria. Since the Yunnan clinical collaboration team was less technically skilled, on 23 October, the Guangdong team sent one of their doctors to help out in the trial. The staff of the Yunnan team returned to their work unit on 6 November. After consulting the 523 Office for permission, the remaining drugs were handed over to the Guangdong team to continue with the trial. According to the records, Liu Pu from the ATCM returned to Beijing on 3 December and gave a detailed and complete report regarding the effects of the drug on the clinical cases observed by the Yunnan and Guangdong teams (Table 3).

Between September and November 1974, the Yunnan team treated three cases: one case of *P. falciparum* malaria and two cases of tertian malaria; Guangdong Academy of Chinese Medicine 523 team treated a total of 18 cases: 14 cases were *P. falciparum* malaria (including three cases who were dangerously ill) and four cases of tertian malaria.

The three cases treated by the Yunnan clinical collaboration team used two doses of 2g for the first day, and two doses of 1.4g for each of the second and third days. The total dosage given to an adult was 4.8g. The fever of one of the *P. falciparum* malaria patients subsided 3.2 hours after consuming the drug. Tests for asexual *Plasmodium* parasites turned negative 31 hours later. For sexual *Plasmodium* parasites, a decrease in number was observed but tests did not turn negative.

Table 3: 18 malaria cases in the clinical trial by the Guangdong Academy of Chinese Medicine, 1974

Treatment	Total dosage (g)	*P. falciparum* malaria cases	Tertian malaria cases	Average time for test for the parasite to become negative (h)	Return of symptoms and *Plasmodium* parasites
Day 1	0.2		1	43.5 for tertian malaria.	Short-term reviews were done for 7 *P. falciparum* cases. 6 cases had symptoms and the *Plasmodium* parasites recur within 8–24 days after medication. 1 case did not have any relapse after 11 days. No further review was done.
	0.3		1		
	0.6		2		
	1.0	1		Tests for 2 cases did not turn negative. Tests for the other 12 cases turned negative after an average of 54h.	
	1.5	1			
	2.0	1			
Day 2	0.9–1.2	3			
	1.5	4			
Day 3	1.5	1			
	2.0	2			
	4.8	1			

For the two cases of tertian malaria, the average time taken for the fever to subside was 13 hours, and the tests for the *Plasmodium* parasites turned negative after an average of 32 hours.

After the clinical trial, the Guangdong Academy of Chinese Medicine 523 team concluded that *huanghaosu* was a fast-acting anti-malarial drug. An initial dosage of 0.3–0.5g was sufficient to quickly control the maturation of the *Plasmodium* parasites. The reason for the early return of the symptoms and parasites might be due to a rapid excretion rate (or the rapid rate the drug was converted to other compounds within the body). As the drug concentration in the blood was not maintained for a sufficiently long period, it was unable to completely eliminate the *Plasmodium* parasites. It was also mentioned for the first time that *huanghaosu* had the characteristics of a highly effective and fast-acting drug that could be used to save patients who were dangerously ill with malaria. It was suggested that an injectable form of *huanghaosu* be developed as soon as possible.

These three institutes used different methods and medicines from different places of origin to extract effective anti-malarial crystals, which were used by different doctors at varying times, locations, and dosages. The clinical trials were carried out separately with differences in their results (Table 4):

The three institutes used different dosages in their clinical tests, with Beijing using a greater quantity of the medicine than Shandong

Table 4: Comparison of results from the clinical trials carried out by the three institutes

Institutes		Beijing	Shandong	Yunnan (including the 2 teams from Yunnan and Guangdong)
Tertian malaria	No. of cases	3	19	6
	No. of successful cases	3	19	6
P. falciparum malaria	No. of cases	5		15
	No. of successful cases	1		13

and Yunnan. However, all three institutes managed to prove the effectiveness of their extracts against tertian malaria, with tests for the *Plasmodium* parasite turning negative in 100% of the cases. However, the results for *P. falciparum* malaria were different. There were no cases of *P. falciparum* malaria in the clinical trial conducted by the Shandong *A. annua* L. collaboration group; ATCM had 5 cases of *P. falciparum* malaria but, for various reasons, only one case was treated successfully; the Yunnan clinical collaboration team and the Guangdong Academy of Chinese Medicine had 15 cases of *P. falciparum* malaria (Guangdong Academy of Chinese Medicine had 14 cases, including three cases of cerebral malaria) and 13 cases were effectively treated. Therefore, speaking solely of efficacy, it was the Guangdong Academy of Chinese Medicine that validated the effectiveness of *A. annua* (*A. annua* L.) extracts against *P. falciparum* malaria.

Once the analysis of the structure, effective dosage and efficacy of artemisinin had been established, a retrospective review of the results of the clinical trials indicated that the results of the clinical trial by the ATCM were not ideal. It was also discovered that cardiotoxicity had been observed which was not present in the clinical results of the other two institutes. Hence, it was suspected that the monomer extracted by the three institutes might not be identical. The author (Li Runhong) interviewed some of the personnel involved with regard to the results of the clinical trials and the cardiotoxicity issue.

Tu Youyou felt that the unsatisfactory clinical results were due to the use of artemisinin tablets which had disintegration problems, and therefore the drugs were later administered in capsule form. All three patients treated using artemisinin capsules achieved full recovery.

Another researcher from the ATCM thought that the extracted crystals were not in the optimal purified state since they were used in clinical trials immediately after extraction. This could be another cause of the unsatisfactory clinical results.

Other work units felt that the clinical trials from 1973 and the laboratory results showed some cardiotoxicity problems. At the same time, the efficacy was deemed unsatisfactory. They felt that even if there were problems with tablet disintegration, at such a high dosage, the results should have been better if artemisinin was really effective.

Li: Can you outline the cardiotoxicity problem with artemisinin?

Respondent: At the time, a Professor Jing recommended pathology and toxicology screening for artemisinin before using it on patients, so we tested it on a dog and examined the pathology and toxicology slices. A teacher, Wei, from the Peking Union Medical College Hospital who was working at the ATCM observed pathological changes in the lung tissues. So Professor Jing insisted that artemisinin was toxic and should not be widely used on patients.

Li: Were there pathological changes found in the dog's heart tissue?

Respondent: Yes, we examined tissue slices from the heart, liver, spleen, lungs and kidneys, but the most obvious pathological changes were seen in the lungs. There were two different views on this phenomenon. Tu Youyou said that symptoms such as shortness of breath should have been present if the pathological changes were serious, hence it was unlikely to be caused by artemisinin. On the other hand, Professor Jing was very cautious and held firmly to the view that artemisinin was toxic.

Li: Before heading to Hainan, were the drugs tested on humans?

Respondent: I cannot remember clearly what happened after the pathology and toxicology screenings. A few healthy people consumed the drug and showed no signs of cardiotoxicity, so we started using the drug in clinical trials. After which, we invited other people to view the dog tissue slices. Gao from the MOH Research Institute, who had returned from the Soviet Union, viewed the tissue slices and immediately declared that the dog had degenerative pathological issues due to age and not due to the drug. After this, we proceeded to use the drug in clinical trials. However, when the crystals were used in the trials, premature contractions of the heart were observed again. As a result, the ATCM sent me back to join the team to determine the toxicity of the drug.

Since only small quantities of artemisinin had been extracted, it was not possible to have tablets made in a factory. Instead, we made the tablets using a smaller machine in the dispensary of the Peking Union Medical College Hospital. Physician Li Chuanjie from the Acupuncture Institute brought Liu Jufu and other technical personnel from the ATCM to conduct the clinical trials in Hainan. It was Liu Jufu who encountered the premature contractions in patients. Her reports made Li Chuanjie very concerned. As he was neither part of the ATCM nor the artemisinin team but was just given temporary responsibility for the clinical trials, he was very cautious and chose to discontinue the trials. By the time the news was transmitted back to us, clinical trials had been carried out on five people. Our superintendent felt that we should not recall everyone from Hainan since there were no conclusive results. He decided to send Zhang Guozhen (later responsible for the business aspects) to Hainan to find out more about the situation. At the time, drug extraction was still on-going, so Zhang Guozhen brought the drug in capsule form in case the trials might be resumed after studying the situation there. However, the situation was never made clear as only Liu Jufu had heard of the premature contractions while Li Chuanjie had not. Since Li Chuanjie had to bear the responsibility, he preferred not to continue with the clinical trials even though he had not heard of any premature contractions himself. Hence, after the capsules were brought over, they only tested them on another three cases, bringing the total number of cases to eight.

Li: So what pharmacology results did you obtain for this drug?
Respondent: No cardiotoxicity was found.

The issue of whether the extracts used by the three institutes were the same was not something questioned only in recent years. According to documentation from 28 February to 1 March 1974, the researchers from the four institutes in Beijing, Shandong, and Yunnan came together for an *A. annua* research forum at the ATCM with 523 Office and relevant leaders from the ATCM. The members who participated in this forum were: Zhang Kui, Zhang Guozhen, Jing Houde, Tu Youyou, Meng Guangrong, Shi Linrong from 523 Office (236), Liang Juzhong from the Yunnan Institute of Materia Medica, Li Guiping from the Shandong Institute of Parasitic Diseases and Wei Zhenxing from the Shandong Insitute of Traditional Chinese Medicine. During the meeting, researchers from each region reported on the progress of their research on *A. annua* (*A. annua* L.). Tu Youyou from the ATCM reported the results from the previous clinical trials and mentioned some problems encountered in extraction. She reported that the effective anti-malarial molecule that was extracted had a molecular formula of $C_{15}H_{22}O_5$ and molecular weight of 282. After using mass spectrometry, ultraviolet, infrared and MRI techniques, artemisinin was confirmed to be a sesquiterpene lactone; Liang Juzhong from Yunnan reported on the effective anti-malarial crystal III extracted from Yunnan *Artermisia annua* L. f. *macrocephala* Pamp. He confirmed the plant taxonomy and reported results of ultraviolet, infrared and elemental analysis. He described the results of experiments on physical traits, pharmacology and toxicology, and stated that no adverse side effects were revealed in the results; Wei Zhenxing from Shandong recounted the history of the research into *A. annua* and described the extraction process. He also commented on the results of the infrared and elemental analysis. Jing Houde from the Beijing Institute of Materia Medica reported on the pharmacology and toxicology aspects and the cardiotoxicity problem. He also commented on the differences between the three institutes' results:

1. A dosage of 25mg/kg was effective in the trials conducted by Shandong, while Beijing used 100mg/kg to be effective;
2. There was a difference of 80 in the C=O base differentiation between Shandong and Yunnan;
3. The melting points were different. Shandong: 150–151°C, Yunnan: 149–150.5°C, Beijing: 155–156°C, (Tu Youyou reported 156–157°C in her report) and whether a Eutectic analysis was carried out;
4. Toxicity: Beijing, heart 200mg/kg; 100mg/kg for intra-gastric administration. In Yunnan, the drug was administered by injection into the abdomen, and the usual dosage was 5 times higher than for oral administration.

Due to these four differences, Jing Houde suggested that the crystals extracted by these three institutes might not be identical. From the remarks made during the meeting, when such a doubt was raised, people of each institute had their own answers. For instance, regarding the dosage issue, Tu Youyou said that a dosage of 50mg/kg from Beijing was able to bring the *Plasmodium* parasite count to nil.

Subsequently, Yunnan reported that the effective dosage they used was lower than 50mg/kg. This showed that the effective dosage of the Yunnan and Shandong trials was lower than that of Beijing. Shi Linrong from the 523 Office said that chemical experiments had to be done under standard conditions. If results were inconsistent, they should be analysed. If any differences were found, researchers should communicate and exchange information. He suggested that each institute could also repeat the experiments to test for toxicity with chloroquine as a control. Tu Youyou suggested that the Shandong researchers should also observe the effects of the crystals on the heart.

According to later experimental results from the Yunnan and Shandong teams, no obvious cardiotoxicity was observed. Regarding the concerns mentioned by Jing Huode, there are no records to indicate that eutectic analysis was carried out, and there is also no concrete evidence that indicate that the crystals extracted by the three institutes were the same compound. However, the crystals were preliminarily regarded to be the same compound. A letter written by Liang Juzhong mentioned that during the meeting, samples from the three institutes underwent infrared analysis and it was concluded that they were the same compound. In the pharmacopoeia today, the artemisinin in use now has a melting point of 150–153°C.

After these research institutes completed their clinical trials, from 1975 onwards, the national and regional 523 Offices launched a national collaboration across multiple systems, work units and specialisms. One of the main aims of this collaboration was to carry out clinical trials. Other than the testing of oral forms of artemisinin, much effort also went into testing other forms of the drug.

Collaboration in the Discovery of Artemisinin

The process of discovering of an effective drug, extracting it in large quantities, and determining its efficacy through clinical trials requires collaboration between many different organisations and disciplines to be successful. Due to limited information, it is not possible to give a detailed account of the collaboration process. An outline is given below based on the information available.

Within work units

Much collaborative teamwork went into the process from the discovery of the efficacy of crude extracts of *A. annua* on malaria-infected mice to the extraction of crystals with 100% efficacy. At the time, research conditions were very different from today's division of labour and detailed job scopes. This was not the way then. In 1973, the ATCM researchers felt their way and experimented through the process of separating the effective pure crystals. There were no specialised equipment or well-established methods for extraction, and members of the research team would make suggestions as to how to proceed. According to the recollection of a participating researcher:

> During that time, very limited quantities of raw materials could be purchased, and many items were unavailable. It was very difficult to buy things as work had not fully resumed. Sometimes, I would personally have to go out to buy the things required. Back then, we were feeling our way through the research work as it was all new to us. Even after the chemistry laboratory was set up, we did not actually perform in-depth work, just simple things. We would usually have discussions before doing the extractions. Although we extracted the crystals using silica gel column chromatography eventually, in reality, many other methods were explored during the research. We divided up the jobs because it was impossible for one person to do everything. There have been great improvements in our ability to obtain crude. In those days, we did not have a defined standard in chemistry and we could only seek direction from pharmacology. If pharmacology indicated that a particular item was more potent, we would focus on that item.

After extracting a small quantity of crystals in the laboratory, many personnel from the ATCM participated in expanded experiments so as to obtain larger quantities for the clinical trials. Another researcher involved in the extraction recalled:

After that, more colleagues joined us and, at the peak, there were up to seven or eight in the team. There were also some people who came to help out temporarily, so there were really a lot of people who participated in this work. At that time, there was a group of students from the cadre training school who had not been assigned to jobs yet, so they helped to do some unskilled work. Later on, other colleagues went over to other places to carry out the extraction, but we did not go as we had a lot of work to complete. If I had gone too, I could not have done the more in-depth work.

Other work units also participated in the process from laboratory to drug manufacturing, but conditions in society then meant that many such attempts were not developed fully. For such specialised work to be successful, cooperation from many people was needed.

Between work units

The work units in Yunnan and Shandong province increased their focus on *A. annua* and managed to obtain *huanghuahaosu* (*huanghaosu*) crystals following the ATCM's confirmation of the efficacy of crude extracts of *A. annua*. Although they learnt about the anti-malarial effect of the *A. annua* extracts at different times and screened plants of the *Artemisia* genus individually, each work unit communicated frequently with the ATCM. During this process, the 523 Office played a very important role. Transportation and communication back then were not as developed as they are now, and it was the 523 Office that organised and held professional meetings to allow research units to be exchange the newest research findings, hence hastening the research progress. However, this later became a cause of disputes.

In 1974, from 28 February to 1 March, the *A. annua* researchers from the four institutes from Beijing, Shandong and Yunnan came together with the 523 Office and relevant leaders from the ATCM for an *A. Annua* Research Forum. During the forum, each work unit presented their conclusions and gave a review of their *A. annua* anti-malarial research. They detailed their research processes, speed of extraction, results of their

的抗疟药，但部分病人有近期复发和有一定的副作用，有效成分提取工艺尚待改进。必须进一步发扬共产主义协作精神，协同作战，加快研究步伐，一抓到底，争取早日做出成绩，为我国和世界劳动人民做出贡献。

经过反复协商和酝酿讨论，一致认为，为加强协作，避免重复，协调任务，加快进度，经商定一九七四年的任务和分工的意见如下：

一、青蒿有效成分的结构鉴定，一九七三年各单位先后都提到有效结晶，初步认为可能是同一个物质，建议由北京中医研究院中药研究所继续与上海有关单位协作，尽快搞清化学结构。

二、青蒿有效结晶提取工艺的改进、药用部分、采收季节和资源调查等，由三个单位结合本地区情况进行研究。

三、临床前药理工作。由云南药物研究所继续进行有效结晶临床前的有关药理研究，北京中医研究院中药研究所继续搞清有效结晶对心脏的影响。

四、临床验证，今年十月前完成150—200人青蒿有效结晶的临床疗效观察（其中恶性疟50例，间日疟100—150例）。临床验证由山东省寄生虫病防治所和云南省疟疾防治研究所负责，中医研究院等有关单位派科研人员参加。

要抓紧临床验证药品落实。由山东省中医药研究所提取150人份，云南药物研究所提取30人份，中医研究院提取50人份有效结晶。

五、青蒿简易制剂的研究。青蒿药源丰富，研制适于基层医院就地生产使用的制剂，是普及与提高相结合，防治疟疾的重要任务。山东省中医药研究所和中医研究院中药研究所于六月底提出简易制剂方案，第三季度进行临床验证。山东、云南现场要对青蒿的各种民间简易用法进行验证。

为加速青蒿的研究，各单位要加强联系，互通情报，及时交流经

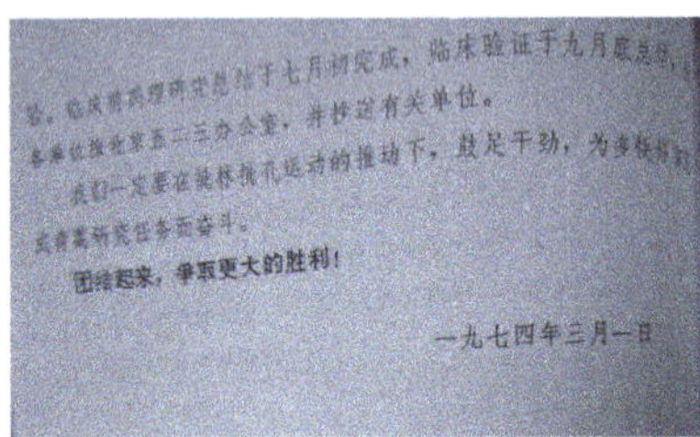

验。临床前药理研究应于七月初完成，临床验证于九月底总结，各单位报北京五二三办公室，并抄送有关单位。

我们一定要在批林批孔运动的推动下，鼓足干劲，为多快好省地完成青蒿研究任务而奋斗。

团结起来，争取更大的胜利！

一九七四年三月一日

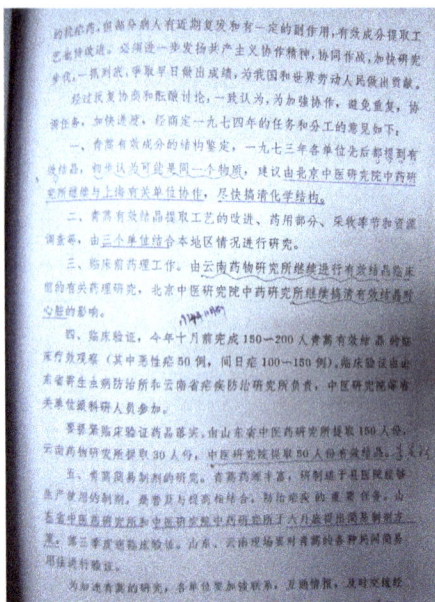

Figure 12: 1974 A. annua research tasks and work distribution

pharmacology and toxicology experiments, information on preliminary clinical trials, and chemical structure analysis. Although there was much repetition, there was also new material. They also made suggestions about future work such as establishing a clearer division of the work to prevent unnecessary duplication. They then drafted a detailed plan for future research tasks and their distribution as shown in Figure 12.

The communications between the main work units and relevant personnel in 1974 has already been described in the previous chapter and will not be further elaborated. After the national meeting in 1975, the number of work units and participating personnel greatly increased. Up till the 1978 A. *annua* meeting, the number of work units that had participated and collaborated in A. *annua* research numbered more than 45. These units used A. *annua* and artemisinin preparations to carry out clinical trials on a total of 6,555 cases. Artemisinin preparations were used in 2,099 cases, of which there were 588 *P. falciparum* malaria cases and 1,511 tertian malaria cases. Of the *P. falciparum* malaria cases, 141 cases were cerebral malaria.

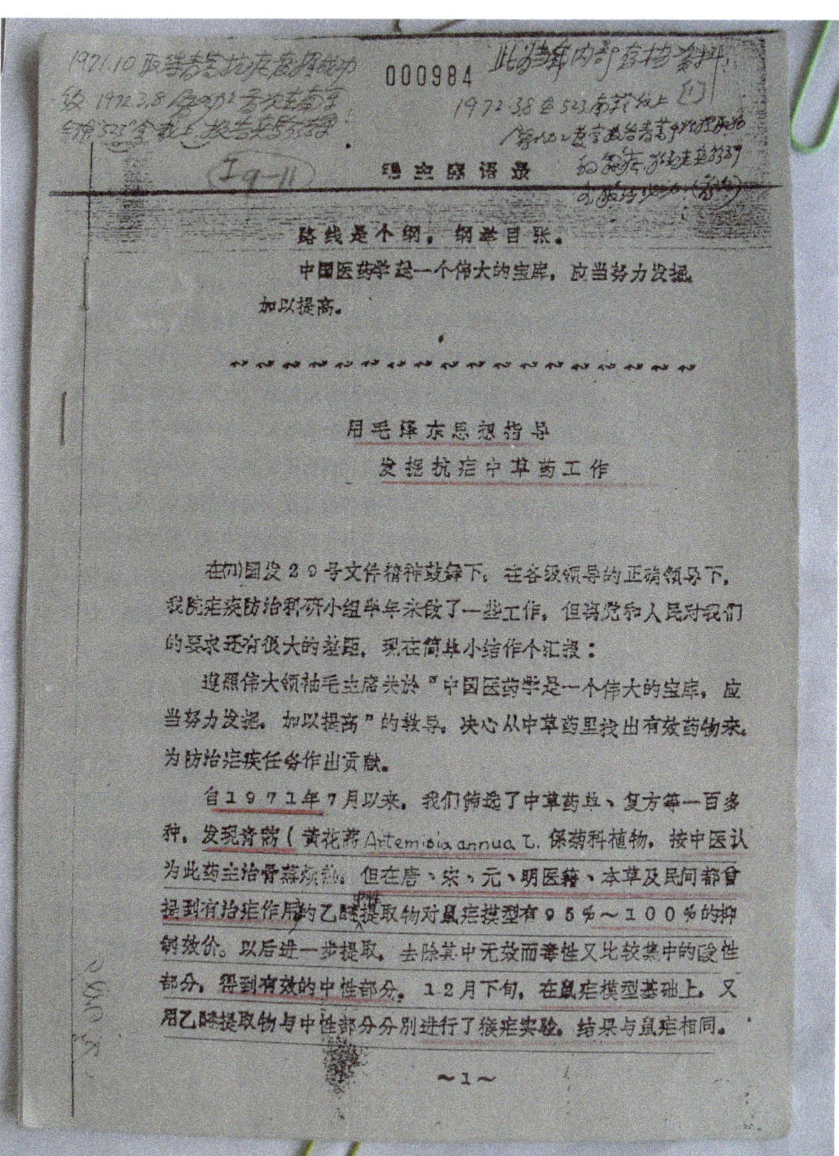

000984

毛主席语录

路线是个纲，纲举目张。

中国医药学是一个伟大的宝库，应当努力发掘，加以提高。

用毛泽东思想指导
发掘抗疟中草药工作

在(71)国发29号文件精神鼓舞下，在各级领导的正确领导下，我院疟疾防治科研小组半年来做了一些工作，但离党和人民对我们的要求还有很大的差距，现在简单小结作个汇报：

遵照伟大领袖毛主席关于"中国医药学是一个伟大的宝库，应当努力发掘，加以提高"的教导，决心从中草药里找出有效药物来，为防治疟疾任务作出贡献。

自1971年7月以来，我们筛选了中草药单、复方等一百多种，发现青蒿（黄花蒿 Artemisia annua L. 係菊科植物，按中医认为此药主治骨蒸烦热，但在唐、宋、元、明医籍、本草及民间都曾提到有治疟作用）的乙醚提取物对鼠疟模型有95%～100%的抑制效价。以后进一步提取，去除其中无效而毒性又比较集中的酸性部分，得到有效的中性部分。12月下旬，在鼠疟模型基础上，又用乙醚提取物与中性部分分别进行了猴疟实验，结果与鼠疟相同。

~1~

The report by Tu Youyou at the Nanjing meeting held on 8 March 1972.

Chapter 3
Scientific Research Collaboration during a Special Historical Period

For me, what is most gratifying is that we finally found a cure for a disease that has plagued hundreds of millions of people in the world every year. Now that it has received international recognition, it has also brought glory to my country.

— Tu Youyou

Several Important Meetings

In an era when politics reigned supreme, scientific research was strongly influenced by political factors. Project 523 was no exception. According to the information available, a total of 50 to 60 meetings, large and small, were held throughout the duration of Project 523. These meetings determined the progress of Project 523 and demonstrate the interaction of politics and science during the Cultural Revolution. This chapter will outline a few of the more significant meetings.

The first meeting at the beginning of the project

In 1967, at the height of the Cultural Revolution, much research work came to a standstill. However, as the mandate for Project 523 came from the highest leader, it was allowed to continue. The Anti-Malaria Drug Research Work Meeting of May 1967 has been described in an earlier chapter and its significance goes without saying.

Further meetings as the project gathered momentum

From 21 to 29 February 1968, the Science and Technology Commission (STC) and General Logistics Department (GLD) jointly held the second national collaboration meeting at the Hangzhou Pingfeng Mountain Workers' Sanatorium. This meeting discussed the further implementation of the plans drawn up during the first national collaboration meeting held in Beijing on 23 May 1967. The meeting also collated the results obtained by each research team since the first meeting and assigned the tasks for the following year. A summary of the meeting's proceedings was then sent to the Central Military Commission (CMC). From then on, every research plan and participating unit of Project 523 was placed on an official footing. In addition to the summary of the meeting, the rules and guidelines discussed during the meeting were also disseminated to the work units involved. Responsibilities were divided among the different departments, regions, work units and specialist teams in order to support the leadership, assign tasks and designate responsibilities.

On 15 January 1969, every province, city and region set up their own Revolutionary Committee. In order to ensure support for anti-malaria drug research from the different Revolutionary Committees, the STC and GLD wrote an *Anti-Malaria Research Report and Request* to the prime minister and the CMC. The report suggested holding a forum in either Beijing or Guangzhou for the personnel in charge from the Revolutionary Committees of each province, city, and region, general logistics departments of each military district, and other work units.

Prime Minister Zhou responded with specific instructions regarding the meeting venue:

> *"Provisional agreement granted. It is better to convene this forum in Guangzhou."*

On 24 January, this request report was signed by Chairman Mao.

On 8 February 1969, a telegraph was received from Prime Minister Zhou. The content of the telegraph:

> *"Notice regarding the convening of the Anti-Malaria Drug Research Forum:*
>
> *With the approval from our great Chairman Mao, the Anti-Malaria Drug Research Forum shall convene in Guangzhou. The STC and GLD will compile a list of personnel who should participate and other related issues. The specific time and venue will be made known at a later date."*

According to the recollection of Shi Linrong, a staff member from the 523 Office:

> *A staff officer who later became one of the commissioners of the GLD first received the telegram. The commissioner phoned director Zhang Jianfang to request that he and the others gather at the Academy of Military Medical Sciences' (AMMS) south gate so that he could show them the telegram.*

According to the recollection of director Zhang Jianfang:

> *Someone from the GLD brought the document signed by Chairman Mao to the AMMS. Everybody was asked to gather at the gate of the Academy to receive the document. We thought Chairman Mao had issued instructions to the Academy, but we did not manage to see anything in the end. I only saw a circle drawn in red and blue ink.*

The 1971 Guangzhou meeting — further strengthening project 523

As the Cultural Revolution intensified, the operations of the STC, regional administrative bodies and research units were interrupted by the revolutionary activities. There were major changes to the leadership and organisation systems. As the newly-appointed personnel might not have been aware of the importance of Project 523, the project encountered some difficulties. Furthermore, the 3-year plan drawn up in 1967 for anti-malaria drug research had come to an end. Hence, on 16 March 1971, the Military Control Commission of the Ministry of Health (MOH), the Ministry of Chemical Industry (MCI), Chinese Academy of Sciences (CAS), and General Logistic Department (GLD) submitted an *Anti-Malaria Research Report and Request* to the State Council and CMC. The report requested changes to the project leadership, appointing the MOH to be the team leader and the GLD to be the assistant team leader, with the office to be still situated at the AMMS. On 15 April 1971, the State Council and the CMC passed the (71) State Document No. 29 to approve the request. On 22 May 1971, the Anti-Malaria Drug Research Forum convened in Guangzhou. During the forum, the six original organisations of the 523 Leadership Group which consisted of Science and Technology and Commission (STC) (leader), GLD (assistant leader), Commission for Science, Technology and Industry for National Defense (COSTIND), MOH, MCI and the CAS were changed to the Ministry of Health (leader), Health Department of the GLD (assistant leader), MCI and the CAS. In addition, the meeting drafted a 5-year plan for anti-malaria drug research for the period 1971 to 1975. One staff member recalled that someone grumbled during the meeting that Project 523 was becoming "ineffective and diffuse", and hoped that the CMC would continue to support the work and strengthen research. On 28 May, the meeting passed a letter of instruction written by Prime Minster Zhou to Xu Jingxian, Deputy Director of the Shanghai Revolutionary Committee. The letter reported that Cambodia's Prince Sihanouk's doctor, General Riche, had offered an anti-malaria prescription to China. The instructions from Prime Minister Zhou were:

> *"Comrade Xie Hua and Wu Jieping, after reading this letter, please pass it to the relevant units of the Academy of Traditional Chinese Medicine (ATCM) and AMMS to carry out further research. Check if it is possible for one or two small teams to carry out trials of this prescription in P. falciparum endemic regions in Hainan and Xishuangbanna, Yunnan. If the prescription is effective, we can supply large quantities of it to support the Indochina battlefield, as they suffer because of malaria."*

On 29 May, the MOH gathered the personnel in charge from the Health Department of the GLD, MOH, ATCM and AMMS to draw up a proposal to carry out trials of this prescription. It was sent to the Prime Minister who replied: "In-principle agreement to do so."

The instructions issued by Prime Minister Zhou were welcomed by the researchers as being of significant support of Project 523's leadership and its implementation. Prior to this meeting, the ATCM had requested to pull out of the research but this was rejected by the MOH. After the meeting, to fulfil the demand from the leaders for a stronger leadership, the Revolutionary Committee and Military Control Commission of the ATCM took steps to strengthen the leadership and research capabilities of the research team, thus ensuring the continuance of the work. Tu Youyou has commented before on the significance of this meeting which resulted in the ATCM anti-malarial research team renewing its efforts. The team evaluated over 200 Chinese medicines and 380 extracted samples before finally focusing on *Artemisia annua* (*A. annua*). Hence, it can be said that the meeting in 1971 had great significance to Project 523 and the rediscovery of the anti-malarial properties of *A. annua*.

The research plan for the following five years was drawn up during the meeting. The meeting also mentioned the anti-malarial prescription provided by Prince Sihanouk's doctor in Cambodia. It was found that this prescription was very similar to the Malaria Drug No. 2 produced by the research centre at the AMMS 236, differing only slightly in the dosage administered.

The 1973 Shanghai meeting — consolidating and intensifying the research

As Vietnam had signed a truce, less urgency was felt towards war preparation; at the same time, the increased difficulties of the scientific research work caused many researchers to lose confidence. On 15 February 1973, the MOH, MCI, COSTIND, and GLD submitted an *Anti-Malaria Drug Research Report* to the Prime Minister in order to report on the progress of the five-year Anti-Malaria Drug Research Work Plan, summarise experiences and the progress of the collaborative research, and further promote the progress of the anti-malarial research work. The document reported on the implementation of the (71) State Document No. 29 passed by the State Council and the CMC, and the progress on the five-year plan. The report also described the clinical trials of the prescription given by Prince Sihanouk's French doctor, General Riche, which the Prime Minister had requested. The report proposed that to meet the country's anti-malaria requirements, three types of anti-malaria medicines should be developed for use in endemic regions of China as well as Vietnam; the anti-malaria drug research work should be included as a key project in the national research programme; an anti-malaria drug research forum should be convened; and detailed procedures should be set out for the final three years of the five-year research plan.

After approval had been received from the State Council, the Shanghai Anti-Malaria Drug Research Forum convened from 28 May to 7 June 1973. This meeting was attended by the personnel in charge from the MOH, the Science and Education group of the State Council, MCI, ATCM and GLD, the leaders and specialists from various provinces, cities, autonomous regions, military districts and work units. Representatives of the Chinese communist party (CPC) Schistosomiasis Leadership Team and the Departments of Commerce of the 13 southern provinces, cities and autonomous regions also attended. A total of 86 people attended the meeting.

The key issues of this meeting were:

1) To emphasise the importance of the project so that there would be no slackening despite the truce signed by Vietnam, and that the anti-malaria work should remain an important project to ensure the health of the military. It was noted that there had been a steady increase in the number of malaria cases among civilians in China in recent years, and that had negative impacts on health, industrial and agricultural production, and war preparedness. Hence, the drug research work remained a major challenge. The message from the meeting was this: "Right now, there should be no thought of stopping; instead we need to make faster progress. We must not weaken but become stronger, and we need to produce results as soon as possible. It is an honour bestowed by Party and people to be entrusted with this project." From this, we can infer that some work units had indeed relaxed, hence the importance of the meeting.
2) To set out detailed procedures for the final three years of the five-year research plan, and to further define, based on the original plan, the mission and requirements of the ten research topics.
3) According to the project No. 321 of the *Relevant Ministries under the State Council's Science and Technology Development Plan 1973*, the anti-malaria drug research work had been included in the research plan of relevant provinces, cities, autonomous regions, departments and work units. It was also suggested that there should be periodic inspections, and that the specialist personnel participating in this project should remain relatively stable. The plan also emphasised that there should be collaboration and coordination between all participating work units.

The April 1975 Chengdu meeting — focusing on artemisinin research

From 28 February to 1 March 1974, researchers from the *A. annua* (*A. annua* L.) research teams from Beijing, Shandong and Yunnan participated in the *A. annua* Research Forum with the leaders from the 523 Office held at the ATCM.

From 14 to 24 April 1975, the Project 523 Traditional Chinese Medicine Forum convened in Chengdu. A total of 62 representatives from the 523 traditional Chinese medicine research teams of Beijing, Shanghai, Jiangsu, Guangdong, Guangxi, Sichuan, Yunnan and Shandong attended the forum. Representatives from relevant work units from Henan, Hunan and Hubei, and several senior Chinese physicians and rural medical personnel were also present. Each research unit reported on the progress of their research. There was a special focus on the traditional Chinese medicine research team from the Guangdong Academy of Chinese Medicine. Over a period of eight years, they had penetrated deep into malaria-endemic rural villages, accumulating experience in treating cerebral malaria and obtaining commendable results. The meeting also mentioned that some research units focused on laboratory research, and had a tendency to work behind closed doors.

The main focus for this meeting was on A. *annua* research, especially on Li Guoqiao and his team from the Guangdong Academy of Chinese Medicine, who reported that the *huanghaosu* obtained from the Yunnan Institute of Materia Medica had been effective in treating 18 cases of malaria. The results obtained were more obvious in *P. falciparum* and cerebral malaria cases. This was a big encouragement to the people in the meeting, especially to the team members of the Chinese traditional herb research team. There were reports about other herbs during the meeting too, for example the anti-malarial properties of *yingzhuasu-A* and *Agrimonia pilosa* Ledeb. Just as described in the book *A Detailed Chronological Record of Project 523 and the Discovery and Development of Qinghaosu* (*Artemisinin*), this was the first meeting at which A. *annua* was the main focus. After the meeting, the research units began to collaborate in further researching the dosage of simple A. *annua* preparations, production techniques, resources, and later, the chemical structure. The meeting also directed that research for artemisinin derivatives should begin.

The 1977 Nanning meeting — preparing for to unveil artemisinin

In May 1977, the National 523 Office convened a Specialist Group Meeting on Integrative Medicine for Malaria Treatment in Nanning,

Guangxi. The meeting summarised and evaluated the research into *A. annua* since the meeting in Chengdu two years before. The meeting specifically detailed the tasks to be completed before the final evaluation of the results obtained. The purpose of this meeting was to prepare for the final evaluation of the research into artemisinin.

The 1978 Yangzhou meeting — appraisal of artemisinin

The meeting was considered to be a summary of the artemisinin research up to that point in time. This was the unveiling of the new Chinese anti-malarial drug to the world, and the first time that a project meeting was attended by the media.

The 1981 Beijing meeting — the winding up of Project 523

From 3 to 6 March 1981, the Anti-Malaria Drug Research Meeting for the leadership group and personnel in charge was held in Beijing. The preparation for this meeting had been carried out in 1980. For example, on 13 June 1980, the National 523 leadership group meeting was held in Beijing. The relevant personnel in charge of the STC, State Administration of Medicine, GLD, MOH, and AMMS attended this meeting. The meeting was hosted by Deputy Minster Huang Shuze of the Ministry of Health and affirmed the Project 523 research work and the results obtained over the past 13 years. The meeting also set out regulations to govern the work after the national and regional Anti-Malaria Drug Research leadership groups and organisations had been dissolved. In August 1980, the four leadership departments, the MOH, STC, State Administration of Medicine and GLD jointly requested that the State Council and the CMC revoke the national anti-malaria research collaboration and to incorporate the anti-malaria drug research project in the plans of relevant organisations in the provinces, cities and states. The meeting in 1981 mainly executed the changes to the Project 523 organisation structure. However, the anti-malaria project remained an important priority and was included in the regular research plan of relevant departments from the states, provinces, cities, autonomous regions and military. In view of the adjustments made to the Project 523 organisation structure, the MOH set up a Malaria Committee within its Medical Sciences Committee. The military also decided that the GLD would organise and plan for an anti-malaria team within the Epidemic Specialist Team in May 1981. On 11 May 1981, representing the Anti-Malaria Drug Research Leadership Team, the four leading departments jointly issued a final document — the National Anti-Malaria Research Forum Summary. Besides the summary, the notice also outlined a general plan for the winding up of Project 523. The regional leadership group administration departments would determine the handling and transfer of the 523 Office documents, technical files, funds and supplies, and also make arrangements with the original work units of the full-time Project 523 personnel who had been away for a long period of time.

During the meetings in 1967 and 1968, clear instructions had been issued regarding the confidentiality of this mission. Especially after the meeting in 1968, it was requested that code numbers be used for all drugs that had undergone clinical trials and whose dosages had been confirmed. In 1972, in order to enable a greater exchange of scientific research results, there was an instruction to sort and compile the anti-malaria research information gathered during the past few years. It was stated that the confidentiality rules should not limit the compilation of this information, and drug names should be used instead of code numbers. If codes were to be used for convenience, the name of the drug should be indicated as well. In November 1972, in a report on the anti-malaria drug research work, it was stated that as long as no national or military secrets (such as the incidence of malaria or the military purposes of this project) were revealed and approval was obtained from the research unit, a researcher was permitted to publish their research in any domestic publication under his own name or in the name of the research unit. From the 1973 Shanghai meeting onwards, it could be observed that Project 523 began to slowly change from a war preparation project to a civilian project. Thereafter, the confidentiality requirement became less stringent. Numerous work units which were not involved in Project 523 originally were invited to collaborate and the publication of research papers was not prohibited. Especially after the first publication of a paper under the name of the artemisinin collaboration group in 1977, many papers regarding the analysis of the structure of artemisinin were published. It was during the National 523 leadership group meeting in 1980 that it was suggested that Project 523 should be incorporated in the national research plan, and this was reported to the State Council and CMC. After the 1981 meeting, Project 523 was formally included in the research plans of the relevant commissions, provinces, cities, autonomous regions and the military.

Collaboration between Personnel and Work Units

From 1967 onwards, large numbers of work units and personnel collaborated in the anti-malaria drug research. Due to space constraints, this section will briefly outline the collaboration between the work units and research personnel. The mode of operation at that time will also be described.

Collaboration between specialist groups

In the earliest plan, the various specialist groups were regarded as independent units. However, as the research progressed, the division became blurred with areas of overlap. Furthermore, as the number of different research groups increased and the research work became more in-depth, the division between groups blurred to the point that they had practically merged. For instance, the initial procedure was that for every new type of drug discovered, the collaboration team would evaluate it according to various criteria (chemical synthesis, pharmacology, clinical use). Drugs that were considered safe, fast-working and long-acting would then be reported to the leadership group for approval. After that, the drug would undergo clinical trials and wider testing in the field. However, there were several research units which handled multiple research topics, such as the Shanghai Pharmaceutical Industry Research Institute, Shanghai Institute of Materia Medica, CAS and AMMS 236. They concurrently handled research on drug synthesis, drug selection, mosquito repellents as well as other topics. Since many of the drugs studied by the traditional Chinese medicine team and mosquito repellents team were derived from plants, there was an overlap between the two teams. For example, large eucalyptus leaves and *A. annua* were studied by both teams. In another instance, the Guangdong Institute of Chinese Medicine was initially a member of the traditional Chinese medicine and acupuncture team, but later moved to the cerebral malaria team. However, as some researchers from the Institute achieved successful results from using *huanghaosu* to treat cerebral malaria, they could also be considered as part of the traditional Chinese medicine team. Later on, as they took part in clinical trials of different formulations of artemisinin, they could also be considered to be part of the field investigation team.

Collaboration across different work units or specialisations

A great deal of effort was devoted to the research and extraction of artemisinin and other drugs. This project differed greatly from laboratory work today where the individual job scopes are clearly defined. In those days, many work units were not fully engaged in the project. When someone was asked to perform a task, he would just join in. For example, many different ideas were tried during the artemisinin extraction process in the ATCM, and the final decision of using silica gel column to obtain more refined crystals was only chosen after trying many other methods. The transition from extraction in a laboratory to production in pharmaceutical factories also involved many personnel. A researcher recalls that a number of students who had just graduated from the cadre training school were assigned to join the extraction work. Similarly, at the Yunnan Institute of Materia Medica, the process from laboratory to factory also involved many different people with no clear delineation of the responsibilities or specialities. During the clinical trials of artemisinin, some of the personnel were from the original research units, but as doctors were required to carry out the trials, more doctors as well as technical staff who had undergone short-term training were co-opted. This was especially so in carrying out clinical trials in the provinces and cities as the number of participating work units and required personnel were very large.

The two diagrams below illustrate the communication flows between different work units during the development phase of Project 523.

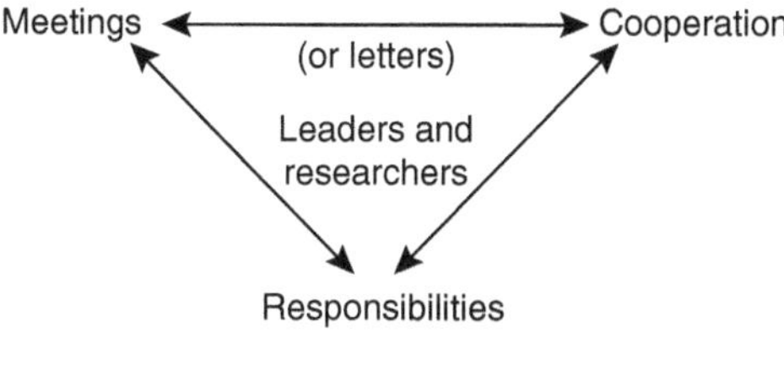

Diagram 1.

Diagram 1 represents the organisation structure at that time. The leadership office would hold meetings to draw up the necessary plans. Responsibilities would be assigned during or after the meetings. Subsequently, the plans would be communicated to the regional 523 Offices or research units. In the plans, some work units were specifically designated as part of the collaboration while others were invited by other work units to collaborate where necessary. After a period of time, a meeting would be held to report on the progress of the work or to emphasise certain issues during the progress of the project. The scale of the meetings could vary. Examples of larger-scale meetings were the National Anti-Malaria Drug Research Work Meetings, the leadership meetings of the National and Regional 523 Offices, and meetings of specialist groups. The smaller-scale meetings were the individual specialist group meetings, such as the *A. annua* Research Forum. According to the information available, there were approximately 50 to 60 meetings of different sizes held from 1967 to 1981.

Diagram 2 represents the outcomes of the organisation structure. During the meetings, participating leaders or researchers from each research unit were required to openly share the progress, experiences, key results and difficulties encountered in their work. The meeting would summarise the status of the research and formulate a plan for future research. After the meeting, the organisers of the meeting would collate the proceedings of the meeting and transmit this "brief report" or "information update" to each participating work unit. The information exchange during the meetings enabled participating personnel to build on the latest research findings and learn from experiences and problems encountered by others. Hence, throughout the progress of the project, these exchanges enabled some teams to advance their research

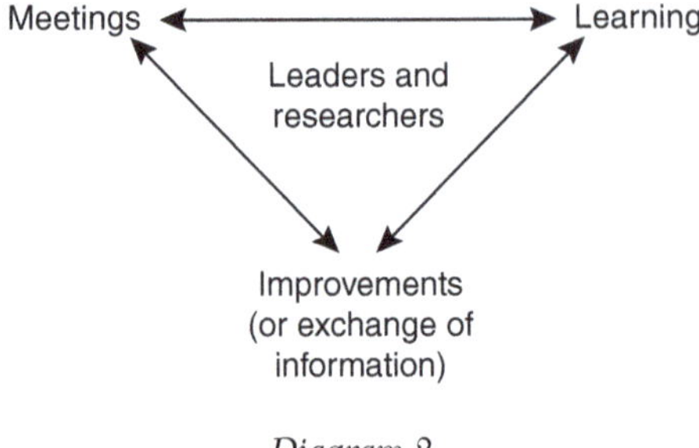

Diagram 2.

while others were influenced to change their focus. For example, many work units screened numerous herbs and discarded those that did not exhibit anti-malarial properties before the traditional Chinese medicine team finally managed to extract an effective active ingredient, artemisinin, from *A. annua*. The acupuncture team, on the other hand, concluded that acupuncture was not a suitable anti-malaria treatment after several years of research and exchanges. Some new research topics were also added to Project 523 as new requirements emerged during the development of the project. For instance, the research on malaria immunity was proposed in the Guangzhou 523 meeting in 1969. After about a decade of research, the research in several areas had achieved international standards.

Most of the people interviewed by the author mentioned that it was an honour, given the environment then, to be able to participate in this project, and they put in their best efforts. They were willing to carry out

Tu Youyou performing an experiment

any task they were assigned regardless of the circumstances. When researchers were unable to attend a meeting to report their results, they would prepare the documents and information for their leaders to share at the meetings.

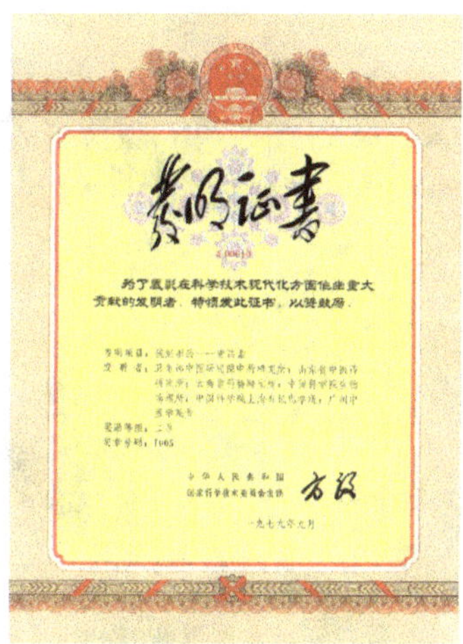

发明证书

为了表彰在科学技术现代化方面作出重大贡献的发明者，特颁发此证书，以资鼓励。

The invention certificate for artemisinin awarded by the Science and Technology Commission

Tu Youyou presenting an academic report

国家科学技术委员会

（96）国科成管　便字第 014　号

证　　明

1992 年，国家科委会同《人民日报》等五家新闻机构组织'92 全国十大科技成就评选活动。国家中医药局推荐参加评选的项目为："抗疟新药－双氢青蒿素及双氢青蒿素片剂的研究"其主要完成单位及主要代表人是：中国中医研究院中药研究所、屠呦呦(详见附件)。

在 1995 年 5 月 30 日全国科技大会闭幕式上，国务委员兼国家科委主任宋健同志在讲话中曾提及我国几位有突出贡献的科技发明家，如王选、王涛等，同时提到青蒿素的发明人屠呦呦同志，对他们的成就给予了肯定。

附 件：

一、《九二全国十大科技成就推荐表》

二、国家中医药管理局科技司修改项目名称便函

国家科委科技成果司

一九九六年五月三十日

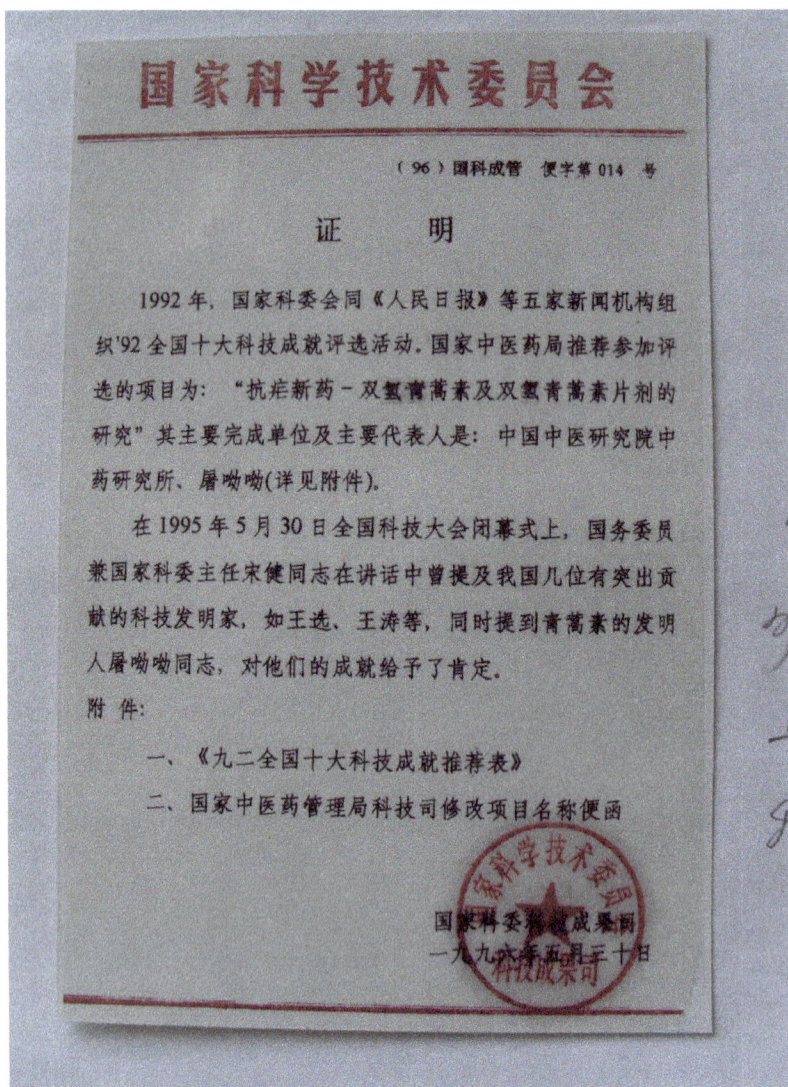

During the National Science and Technology meeting in 1995, Song Jian, director of the Science and Technology Commission and State Councillor, affirmed the results achieved by Tu Youyou

Chapter 4
Artemisinin's History and Future

My formal education at Beijing University, further studies, and knowledge of both Chinese and western medicine were all provided by my country. No matter what my country requires of me, I will strive to do it well.

— Tu Youyou

From the Past to the Present

By Zeng Qingping

Artemisinin is truly "China's wonder drug"! It has cured patients infected with often-fatal cerebral malaria, and has also been effective against chloroquine-resistant malaria in patients on the verge of death. As the mortality rate of malaria in Africa, South America and Southeast Asian countries is comparable to that of AIDS, artemisinin is a truly life-saving drug in malaria endemic regions. Although malaria is only prevalent in the south-western border regions of China, the increase in the number of mosquitoes as a result of global warming has increased the risk of a resurgence of malaria in China. As the Chinese population is large and the likelihood of infection great, artemisinin will be responsible for saving the lives of many malaria patients.

What exactly is the substance "artemisinin"? Where did it come from? Is it unique to China? What lies in the future for artemisinin? What interesting anecdotes are there in the history of its discovery? This section will give a brief introduction to artemisinin, including some information that has been generally unknown up till now. Following that are more details about the people and events in the history of artemisinin.

The sources of artemisinin

Artemisia annua: the only natural source of artemisinin

In chemical terms, artemisinin belongs to the class of sesquiterpene compounds. The peroxide bridge in the molecule is a unique characteristic of artemisinin and it is this that is responsible for its anti-malarial properties and other functions. The only natural source of artemisinin is the aromatic oil in the fine glandular hair of the leaves of the medicinal plant *Artemesia annua* (*A. annua*, also known as *huanghuahao*). To *A. annua* itself, artemisinin is toxic, and this is perhaps the reason why it is isolated in the glandular hair.

Why does *A. annua* synthesise artemisinin? Why is artemisinin produced only by *A. annua*? There is no clear answer to these questions, but evidence shows that artemisinin might play a role in helping the plant resist diseases, worms and stress. Natural disasters such as diseases, worms, heat, cold, drought and floods trigger the cells of *A. annua* to produce large quantities of harmful reactive oxygen free radicals. Dihydroartemisinic acid combines with the reactive oxygen free radicals to create the peroxide bridge in the artemisinin molecule. In another words, artemisinin is the compound which "traps" the reactive oxygen free radicals.

Wild *A. annua* can be found all over the world but the artemisinin content is generally very low. However, the artemisinin levels in *A. annua* in the Yunnan, Guizhou and Sichuan provinces of China are higher. Hence, although *A. annua* can be found in other parts of the world, artemisinin is uniquely a product of China. Records from ancient times found in the Mawangdui Han tombs in Changsha, Hunan note the use of *A. annua* for treating diseases. In later times, *A. annua* was used specifically for the treatment of the fever, chills and shivering caused by malaria. According to the instructions in the *Handbook of Prescriptions for Emergencies* written by Ge Hong during the Eastern Jin dynasty, wild whole *A. annua* was to be immersed in water, then the juice wrung out and drunk to lower the fever.

Artemisinin is a fat-soluble substance and is insoluble in water. Its solubility in blood is also very low, and the low plasma concentration of artemisinin in blood makes artemisinin less effective. To increase the water solubility of artemisinin, the artemisinin molecule have been modified by adding various hydrophilic groups to produce more effective semi-synthetic artemisinin derivatives. Examples of these are artesunate, dihydroartemisinin, artemether and arteether.

The industrialisation of artemisinin production

In 2003, researchers from the University of California, Berkeley synthesised the first precursor of artemisinin, artemisinic acid, using genetically engineered E. coli. In 2006, they synthesised more precursors of artemisinin, including artemisinic acid and dihydroartemisinic acid, using genetically engineered yeast. However, they were unable to synthesise artemisinin. This was because the microorganisms did not produce the reactive oxygen free radicals or the fatty environment needed for the transformation of these precursors into artemisinin.

Nevertheless, American researchers found another way to sythesise artemisinin, which was to put dihydroartemisinic acid synthesised from yeast through chemical catalysis. Artemisinin synthesised from yeast cost US$250-$400 per kilogram whereas the cost of extracting artemisinin from *A. annua* could range from US$120 per kilogram to as high as US$1200 per kilogram. From the economic point of view, using engineered microorganisms to produce artemisinin is highly efficient, economical and practical.

Artemisinin from yeast is currently produced commercially at an Italian factory set up by Sanofi. In 2013, the production of artemisinin reached 25 tons, increasing to 55-60 tons in 2014. In August 2014, the first batch of 17 million courses of artemisinin was shipped to almost half the countries of Africa and saved a vast number of malaria patients. Industrial production of artemisinin has raised its first challenge to the agricultural production of artemisinin. In future, artemisinin produced in China may well become redundant.

The story behind the discovery of artemisinin

The birth of artemisinin from traditional Chinese medicine

Plasmodium falciparum (*P. falciparum*) malaria has been endemic in south China since ancient times and was referred to in ancient writings as "toxic miasma". The use of aqueous extracts of *A. annua* to treat malaria was widespread among the common people. However, it was not known to the scientific community and even less understood by foreign scientists. Also, since *A. annua* had always been regarded as a Chinese medicine, students of western medicine did not have opportunities to learn about *A. annua*. This made it difficult to make the link between *A. annua* and a potential anti-malarial drug.

Although Tu Youyou was trained in western medicine, she had also attended short training courses in traditional Chinese medicine and did not look down on traditional Chinese medical remedies. In addition, during the Cultural Revolution, China lacked doctors and modern drugs, so looking to Chinese herbs as a source of new drugs was the only option. During her investigation in rural areas, Tu Youyou learned that *A. annua* was a common prescription for the treatment of malaria. With her training as a pharmacologist and experience in pharmaceutical chemistry, she immediately suspected that *A. annua* contained some anti-malarial component which she determined to extract.

However, when Tu Youyou used ethanol to extract the anti-malarial component from *A. annua* at high temperatures, she found that the extract performed unsatisfactorily in animal tests. As she continued to experiment with different methods of extraction, she also searched for answers in traditional Chinese medicine texts. Her efforts paid off. She found inspiration in the *Handbook of Prescriptions for Emergencies* by Ge Hong of the Eastern Jin dynasty. Considering the instruction to "immerse in water" and "wring out the juice", it suggested to her that the anti-malarial component might be deactivated by high temperatures, and she deduced that the reason for the failures of the earlier extracts was the high boiling point of ethanol. She decided to try using ether, with its lower boiling point, for the extraction. After switching from high temperature extraction with ethanol to low temperature extraction with ether, she was able to successfully separate an *A. annua* extract which was highly effective against malaria. Subsequently, the molecular structure of this extract was successfully analysed and it was named artemisinin.

(2) The catalyst for artemisinin: aiding Vietnam against America

Thirty years ago, medical conditions and technology were poor in China. So how was it possible that China developed a world-class anti-malarial drug? The 1960s to 1970s was the period of China's Cultural Revolution. Schools stopped lessons, factories ceased production, social life was affected, and the normal operations of research organisations were also interrupted. How was it possible that artemisinin, a scientific achievement of such great significance to humanity, could be produced during this extraordinary historical period?

The launch of the large-scale anti-malaria research stemmed from a secret mission to aid Vietnam against America. In 1964, the American army invaded Vietnam and the Vietnam War began. However, the American army was not able to defeat Vietnam and was bogged down by the war for a long time. Vietnamese military and civilians continued their strong resistance against the Americans in order to save their country. The weather in Vietnam, located as it was in the tropical Indo-Chinese peninsula, was hot and humid. Mosquitoes bred throughout the year in this natural habitat, and *P. falciparum* malaria was prevalent all year round. Both the American and Vietnamese troops suffered greatly from malaria.

Among the American troops stationed in Vietnam in 1965, the incidence of malaria was as high as 50%. During a war operation from Pleiku in Vietnam to the Cambodia border regions, the incidence of malaria in the American troops reached 20%. Even more seriously, within a two-month period, the incidence of malaria in some military units reached 100%. Although statistics are incomplete, the attrition rate of American troops caused by malaria was four to five times that of combat-related attrition. During the four years from 1967 to 1970, the American army had an attrition of 800 thousand soldiers due to malaria, and the actual figure was actually far beyond this number.

Similarly, the Vietnamese troops which moved south from the northern part of Vietnam were severely affected by malaria. Due to the incessant bombings and blockades of the American army, these northern troops had to march more than a month to reach the southern war region. As many soldiers were infected with malaria and needed to be sent to the rear for urgent treatment, only two troops of the original regiment were fit for battle when they arrived.

There was severe drug resistance to some of the anti-malarial drugs in use at the time, such as chloroquine, pyrimethamine, proguanil and atabrine. As a result, the recovery rate was very poor. At the same time, cerebral malaria, the most dangerous form of malaria, had a very high mortality rate. The side that was able to obtain an anti-malarial drug that was highly effective, fast-acting and did not create drug resistance would be the one to win the war.

To find a solution to this difficult problem, the American army set up a specialised malaria committee, greatly increased the funds for malaria research, and gathered tens of organisations to join in the research for a new anti-malarial drug. Institutions such as the Walter Reed Army Institute of Research, the Naval Medical Research Institute and other military and civilian specialists were sent to the Vietnam battlefield to carry out medical and epidemiological investigations and to work on developing an anti-malarial drug. At the same time, America collaborated with research institutes in England, France and Australia and the large pharmaceutical factories in Europe to develop new anti-malarial drugs and prodrugs. The aim was to provide 30 types of prodrugs each year for clinical trials.

By 1972, the Walter Reed Army Institute of Research had made a preliminary selection of 214,000 compounds. Nevertheless, they were not able to find a good anti-malarial drug despite having such a large repository of potential drugs. The only effective anti-malarial monomer they identified was mefloquine. However, mefloquine had adverse side effects and, like chloroquine, it triggered drug resistance easily. It was also not as effective as other new anti-malarial drugs that China had researched and was definitely not comparable to artemisinin.

Vietnam sought China's assistance to resist the American invasion and solve the problem of malaria in the military. The Chinese leader acceded to the request to help look for an effective anti-malarial drug, leading to the launch of the secret emergency project to find and develop anti-malarial drugs. The medical science and technology capabilities of the entire nation were organised into a huge research collaboration to aid the Vietnamese army against America on this war front.

National mobilisation in the discovery of artemisinin

In 1967, China was experiencing the turbulence of the Cultural Revolution and almost all projects had come to a standstill. It was evident that the research capabilities of the military alone would not be equal to the urgent and challenging task of identifying a new anti-malarial drug within a short time frame. As a result, civilian research capabilities were mobilised to form a military-civilian collaboration to provide aid to Vietnam as quickly as possible. The General Logistics Department (GLD) of the People's Liberation Army invited the Science and Technology Commission (STC), Ministry of Health (MOH),

Ministry of Chemical Industry (MCI), Commission for Science, Technology and Industry for National Defense (COSTIND), Chinese Academy of Sciences (CAS) and the China National Pharmaceutical Industry Corporation to commit their research, medical, education and drug production work units to this collaboration. This would be their role in the honourable revolutionary mission.

In order to find a solution to drug-resistant malaria in the tropics, the Academy of Military Medical Sciences (AMMS) drafted a three-year research plan. After preparation, discussion and audit by the leading departments, on 23 May 1967 in Beijing, the STC held a national collaboration meeting attended by the relevant ministries and commissions, the provinces, cities and autonomous districts under the Central Military Commission, military leaders and associated departments. The meeting discussed and drew up a three-year anti-malaria drug research plan. This is the origin of the famous "523 Meeting" and "Project 523" names.

Project 523 lasted for 13 years, involving over 60 research units from all over the country as well as 500-600 regular workers. Including those who were replaced partway through the project, the total number of participants was as high as 2000-3000 people. In the course of three major meetings over two years, the Guangdong, Jiangsu and Sichuan regions used artemisinin and extracts of *A. annua* in the treatment of 2,000 cases of malaria. Of the 800 cases treated with artemisinin, 100% were cured. Of the 1,200 cases which used *A. annua* extracts, the efficacy rate was 90%. The new drug artemisinin was invented in 1975 and received national certification in 1979.

Health Benefits of Artemisinin and Future Research

By Zeng Qingping

Artemisinin's greatest contribution to the health of mankind was in combating chloroquine-resistant malaria. It is a unique molecule which has a peroxide bridge structure, and may also have potential for use against obesity, aging, inflammation, cancer and bacterial infections. If

Tu Youyou and other seniors in artemisinin research are pioneers in the field of research, then research into further uses of artemisinin can be significant in inspiring later generations through the example of their predecessors.

The peroxide bridge structure of artemisinin makes it effective in the prevention and treatment of many different diseases. Artemisinin transforms into carbon-centred free radicals in the body which can combine quickly with the protein or enzyme of the heme prosthetic group and, through obstructing the inter-conversion of ferric and ferrous ions, inhibit the activities of the proteins or enzymes and induce the expression of the related genes to carry out its therapeutic effect.

The most important heme enzymes in cells are the nitric oxide synthases (NOS) of which there are three different types in higher animals and humans. The three types are the endothelial NOS (eNOS) which maintain the normal functions of the blood vessels, the neuronal NOS (nNOS) which maintain the normal functions of the nerves, and the inducible NOS (iNOS) which are responsible for inducing inflammation to kill invading pathogens.

The concentration of nitric oxide synthesised by eNOS and nNOS is lower while that synthesised by iNOS is higher. Lower concentrations of nitric oxide help maintain normal physiological functions and are called physiological levels of nitric oxide. High concentrations of nitric oxide can cause pathological injury to tissues, and are called pathological levels of nitric oxide. Physiological nitric oxide can have slimming and anti-aging effects while pathological nitric oxide can induce rheumatoid arthritis.

The synthesis of nitric oxide, regardless of the source, gradually decreases with age. The advantage of this is that the rheumatoid arthritis often tends to stabilise in the elderly, but the disadvantage is that nitric oxide aggravates visceral obesity, hastens the aging process and increases the incidence of geriatric diseases. Therefore, the ability to flexibly adjust the levels of eNOS, nNOS and iNOS is the key to health.

As a highly effective and life-saving treatment for malaria, artemisinin should receive the highest level of protection. There are some who feel that artemisinin should not be lightly used for the treatment of any diseases other than malaria unless there is no alternative drug. This is to prevent the malaria parasite losing its sensitivity to artemisinin, and to safeguard against a future where there may not be any suitable drug against multi-drug resistant malaria.

Artemisinin has now been used as an anti-malarial drug for more than 30 years and there has been no clear evidence that it can induce drug-resistant genetic mutations in malaria parasites. This could be due to the anti-malarial mechanism of artemisinin being multi-targeted, highly efficient, and of low toxicity. The evidence from molecular pharmacology indicates that artemisinin can be appropriately used in the treatment of diseases other than malaria without excessive concern that its efficacy against malaria will decrease or even disappear.

Nevertheless, the use of artemisinin should be limited to a single course and in combination with other drugs. Limiting the use of artemisinin to a single course will not allow any opportunity for the induction of antioxidant enzymes, and the use in combination with other drugs can overcome artemisinin's shortcomings of being short-acting and with a high relapse rate, as the long-lasting and low relapse characteristics of other anti-malarial drugs can come into play. This will reduce the risk of drug resistance to artemisinin.

Anti-malarial effect of artemisinin

Malaria is a parasitic disease that is propagated through mosquitoes. Among the types of malaria, *P. falciparum* malaria which can lead to cerebral malaria has the highest mortality rate. Malaria is endemic in tropical and sub-tropical regions including Sub-Saharan Africa, Southeast Asia and Latin America, with Africa being the most severely affected. According to World Health Organization (WHO) statistics, there were 198 million cases of malaria worldwide in 2013, causing 850,000 deaths, of which 90% were in Africa.

Chloroquine drug resistance

Before artemisinin was discovered, chloroquine was the main drug used in the treatment of malaria. Chloroquine is semi-synthesised from quinine which is a 4-quinoline compound extracted from the bark of the cinchona tree. In 1934, the German company Bayer had already produced chloroquine, but due to its high toxicity, it was thought unsuitable for human use and was set aside for many years. It was not until the Second World War that the American government conducted clinical trials of chloroquine for the treatment of malaria, and officially approved the use of chloroquine as a preventive drug against malaria in 1947.

However, after a few years of chloroquine use, the malaria parasite developed resistance to the drug. Chloroquine-resistant malaria first appeared after 1950 in Southeast Asia and South America, and by 1980, chloroquine-resistant malaria could be found all over the world.

Artemisinin and chloroquine-resistant malaria

At the time when mankind was facing the crisis brought about by malaria, China's wonder drug, artemisinin, with its unique anti-malarial mechanism, was discovered. After the effectiveness of artemisinin was recognised, the Chinese government unselfishly recommended it to all malaria-plagued regions in the world, and especially to the African regions where malaria was endemic. Today, the WHO recommends the use of artemisinin-based combination therapy against malaria. Artemisinin-based drugs have become the standard treatment for malaria, and artemisinin and its related drugs have been added to the list of standard drugs.

Countless lives have been saved since the introduction of artemisinin to treat malaria. Since artemisinin became the first-line drug in the treatment of malaria in the late 1990s, it has effectively reversed the shrinking of the local population in the poorest regions of the world by saving the lives of those infected by malaria. Malaria affects the health of 200 million people every year, and the use of artemisinin has lowered the mortality rate by 20% (overall) to 30% (among children). In the malaria endemic regions of Africa alone, artemisinin saves the lives of 100,000 people each year.

Without artemisinin, not only would many malaria patients die, the spread of malaria to all parts of the world would cause loss of life on an even greater scale. The malaria parasite, now resistant to chloroquine and other traditional anti-malarial drugs, would spread malaria on the global scale. In a world without artemisinin, the threat that malaria brings to human life would be devastating.

The anti-malarial effect of artemisinin's peroxide bridge structure

Although the exact mechanism of how artemisinin works against malaria is still unknown, it is clear that the formation of free radicals because of the peroxide bridge structure of artemisinin is crucial in producing its anti-malarial effect. In the past, it was thought that artemisinin worked mainly through the malarial pigment produced when *Plasmodium* engulfed haemoglobin, as the chelated iron in the malarial pigment would cause the breakdown of the peroxide bridge, thereby forming a large number of free oxygen radicals which kill the *Plasmodium* parasite. Another theory is that artemisinin may disrupt the cell's redox cycle as artemisinin has been found to inhibit the glutathione S-transferase on the digestive vacuoles of *Plasmodium*. Recently, a new hypothesis has suggested that free oxygen radicals might inhibit *Plasmodium's* ATPase (PfATP6), thereby blocking the energy metabolism process and causing its death.

Some evidence which confirms that PfATP6 is the target of artemisinin: artemisinin can also inhibit the calcium ATPase (SERCA) of mammals, the mutation of PfATP6 and SERCA regulates *Plasmodium's* sensitivity to artemisinin, and that after the liquefaction of PfATP6 and SERCA with detergent, both became insensitive to artemisinin.

Drug resistance in *Plasmodium* to artemisinin-based drugs

Despite clinical observations that some malaria patients do not respond to artemisinin and artemisinin-based drugs, there has been no confirmation to date that there is artemisinin resistance in *Plasmodium*. However, after artemisinin was administered as a single drug for 30 years, clinical evidence of the first case of artemisinin resistance was observed in Southeast Asia in 2008. There were also reports of artemisinin resistance in Thailand in 2012 and in Cambodia, Vietnam and Myanmar in 2014. As of now, it is still uncertain whether this artemisinin resistance originated from the cells of the host or the malaria parasite, or both. It is also unclear whether the resistance is due to genetic mutations or has been induced.

Theoretically, like all other anti-malarial drugs, there is a risk that artemisinin may also induce resistance in *Plasmodium*. However, although artemisinin has been in used for years, only sporadic cases of artemisinin-resistant malaria have been observed. This implies that the drug resistance could have been induced by the *Plasmodium* antioxidant enzymes and not because of genetic mutation. As mature erythrocytes do not have nuclei, it is speculated that the drug resistance induced by genetic expression can only occur in the body of *Plasmodium* and not in erythrocytes.

Artemisinin can stimulate reactive oxygen free radicals which in turn induce the synthesis of antioxidant enzymes and antioxidants in *Plasmodium*, and can lower the sensitivity of *Plasmodium* toward artemisinin and allow *Plasmodium* to hide itself inside erythrocytes to avoid being attacked by the reactive oxygen free radicals. On the other hand, antioxidant enzyme inhibitors can restore the sensitivity of *Plasmodium* toward artemisinin. This explains why although artemisinin can induce drug resistance, drug resistance is not easily induced due to its fast action.

As long as the prolonged and repeated use of artemisinin is avoided, antioxidant enzymes will not be produced and drug resistance will not be easily induced. At the same time, if pro-oxidants or antioxidant enzyme inhibitors are administered with artemisinin, it will be even less likely for *Plasmodium* to develop artemisinin resistance. Nevertheless, to be on the safe side, the WHO has repeatedly emphasised that artemisinin should not be used alone. In future, not only is it necessary to continue the use of other long-acting anti-malarial drugs together with artemisinin, the use of pro-oxidants or antioxidant enzyme inhibitors together with artemisinin should also be considered.

Artemisinin and weight loss

When a person is infected by a large number of pathogens, or a small number of pathogens has invaded the human body and is multiplying quickly, acute inflammation occurs which eventually eliminates the pathogens. One such example is ordinary influenza caused by the influenza virus. However, when the number of pathogens is very low, or the pathogens are able to inhibit the immune system, chronic inflammation occurs and the immune system may not be able to completely eliminate the pathogens. One such example is tuberculosis caused by *Mycobacterium tuberculosis*.

Interestingly, when intestinal microflora suffer ecological disturbance, the overgrowth of some proteoglycan organosulfate degraded bacteria will cause the intestinal wall to be completely damaged, and endotoxins of gram-negative bacteria will leak through the intestinal wall into the bloodstream to cause systemic chronic low-grade inflammation. Through inducing iNOS to stimulate the synthesis of a large amount of nitric oxide, and by competing for arginine, the precursor of nitric oxide, the inflammation is able to repress the activities of eNOS and nNOS, thereby destroying the ability of the mitochondria to regenerate.

Adipose tissues can be categorized by their morphology into brown adipose tissues or white adipose tissues. The former has abundant mitochondria and well-distributed capillaries while the latter has few mitochondria and capillaries. The reduction in the number of mitochondria because of inflammation can cause brown adipose tissue to be converted to white adipose tissue which are not able to carry out aerobic oxidation and energy dissipation. This results in adipogenesis without adipolysis, which eventually leads to overweight and obesity (Figure 13).

Artemisinin acts on the respiratory chain to promote the regeneration of mitochondria, thereby causing adipose tissues to turn brown and accelerating the aerobic breakdown of fats. However, as artemisinin cannot selectively inhibit iNOS, its slimming effect is low. Only when artemisinin is used together with antibiotics (anti-bacterial) and immunosuppressants (anti-inflammation) can it lead to significant anti-obesity effects.

The latest research reveals that obesity might not be a disease. "Healthy obesity" which is the simple nutritional storage of subcutaneous fat is non-inflammatory and does not trigger insulin resistance or lead to type II diabetes. In contrast, "unhealthy obesity" caused by an imbalance in intestinal microflora which leads to visceral storage of fat (belly fat) is inflammatory and can trigger insulin resistance and lead to diabetes. Clearly, only those with unhealthy obesity need to slim down.

Anti-aging effect of artemisinin

The longevity of a person depends on the "Hayflick limit", or the number of times cells can divide. The main reason for aging is the

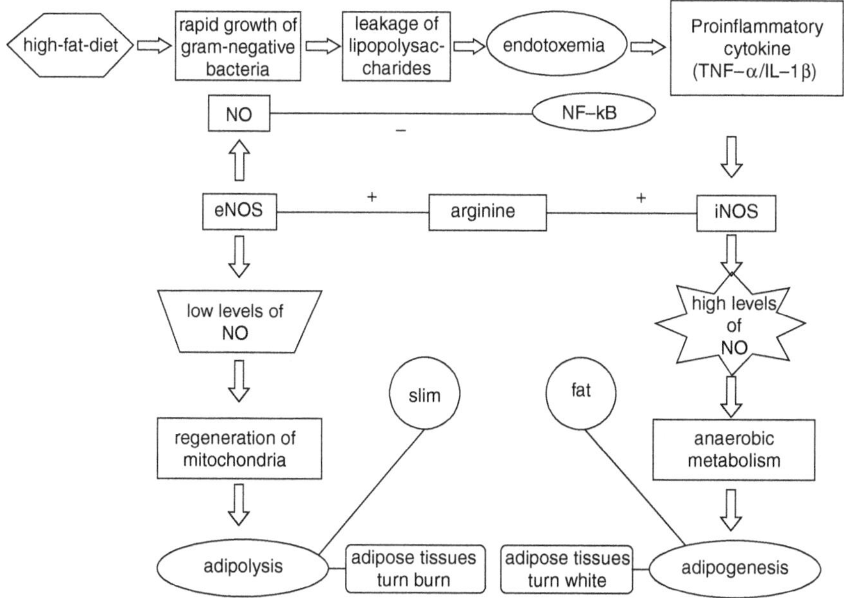

+ : activation; – : inhibition; TNF–α : tumor necrosis factor α; IL–1 β: interleukin–1 β;
NF–kB: nuclear factor kB; eNOS: endothelial nitric oxide synthase; iNOS: inducible nitric oxide synthase.

Figure 13: Artemisinin and weight loss

weakening of cell division activities as a result of the shortening of the chromosome telomere due to the destruction by reactive oxygen free radicals. Reactive oxygen free radicals are formed in mitochondria of cells, and the function of mitochondria decides the fate of cells. If mitochondria function abnormally to produce more reactive oxygen free radicals, these will hasten the process of cell aging. Hence, the main culprit of aging could be dysfunctional mitochondria.

Caloric restriction through dieting can activate eNOS and nNOS, thereby stimulating the mitochondria to proliferate and increasing the anti-oxidative abilities of cells. At the same time, it inhibits the anabolism pathway and activates the catabolism pathway so that even in the state of starvation or low food intake, cells can still rely on autophagy of cell constituents to maintain their basic metabolic functions. As metabolic activities are reduced, the rate of production of reactive oxygen free radicals will also be greatly reduced. This will slow the telomere shortening process and cell aging, and increase longevity.

Artemisinin can simulate the anti-aging effect of dieting as it can also induce eNOS and nNOS, hence stimulating the proliferation of mitochondria. This will reduce metabolism and lower the speed of telomere shortening, thereby achieving an anti-aging effect. The concentration of artemisinin needed to produce this anti-aging effect is extremely low. In yeast, it is only 0.1–0.5 mmol, and in mice it is only 260 mmol. This phenomenon of using a low concentration of a toxic substance to achieve a favourable response is known as "hormesis".

Lastly, an interesting phenomenon is that 2,4-nitrophenol, metformin, resveratrol, and rapamune also have anti-aging effects similar to artemisinin. The secret lies in their abilities to induce eNOS and nNOS either directly or indirectly which stimulates the synthesis of physiological nitric oxide, and also regulate the metabolism through increasing the number of mitochondria. Therefore, nitric oxide has been called the "fountain of youth" and mitochondria the source of the "fountain of youth" (Figure 14).

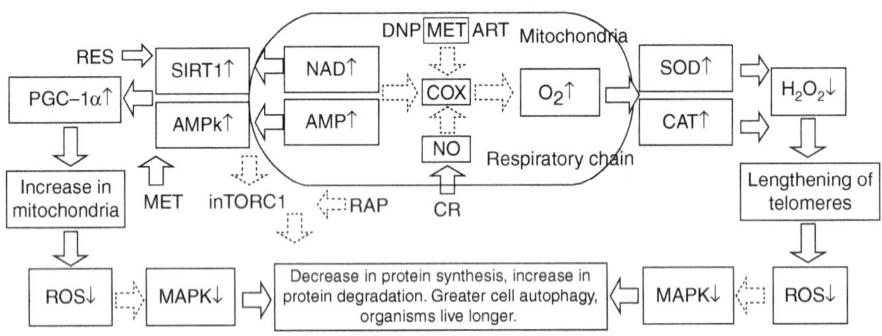

Figure 14 Anti-aging effect of artemisinin

[Solid-line arrow ⇨] activation; [Dotted-line arrows ⇨] inhibition; [Up arrows ↑] increase; [Down arrows ↓] decrease

AMP: Adenosine monophosphate; AMPK: AMP-activated protein kinase; ART: Artemisinin; ATP: Adenosine triphosphate; CAT: Catalase; COX: Cyclooxygenase; CR: Calorie restriction; DNP: Dinitrophenol; H_2O_2: Hydrogen peroxide; MAPK: Mitogen-activated protein kinase; MET: Metformin; mTORC1: Mammalian target of rapamycin complex 1: NAD+: Nicotinamide adenine dinucleotide (oxidized form); NADH: Nicotinamide adenine dinucleotide (reduced form); NO: Nitric Oxide; O_2-: Superoxide anion; PGC-1α: Peroxisome proliferator-activated receptor gamma coactivator 1-alpha; RAP: Rapamune; RES: Resveratrol; ROS: Reactive oxygen species; SIRT1: Sirtuin 1; SOD: Superoxide dismutase

Anti-inflammatory effect of artemisinin

In the past, the cause of rheumatoid arthritis was not understood, but it is now believed that long-term infection resulting in chronic inflammation is the culprit, and the immune system plays a leading role in the process. Hence, it is reasonable to term rheumatoid arthritis as an auto-immune and inflammatory disease.

When pathogens invade the body, other than forming antibodies (humoral immunity) and secreting cytokines (cellular immunity), the cells of the immune system also synthesise large amounts of nitric oxide to assist in the killing of the pathogens. After inflammation occurs, cytokines will induce the activation of iNOS, resulting in an excess of nitric oxide which causes hypoxia in local tissue.

Nitric oxide can trigger cells to enter a state of hypoxia because it competes with oxygen to bind with haemoglobin and myoglobin, causing them to be unable to carry oxygen to cells and tissues in the body. At the same time, nitric oxide can pass through cell membranes and enter mitochondria to combine with cytochrome c oxidase to arrest the electron transfers in the upper respiratory chain. The mitochondria become dysfunctional and aggravate the damage caused by hypoxia.

What are the consequences of hypoxia? Hypoxia stimulates the formation of new capillaries and increases the synthesis of blood components. Inflammatory substances are thus brought to the areas suffering from hypoxia. This eventually causes localised redness, swelling and pain, and damage to cartilage and bones. Joints are organs with relative hypoxia. The oxygen pressure at the synovial surface of the joints is only 6% that of other tissues while the oxygen pressure in the synovial fluid is only 1%. This is why joints bear the brunt of the tissue injury caused by hypoxia.

Artemisinin can effectively inhibit the activity of iNOS and prevent the occurrence of nitric oxide-induced hypoxia, allowing joint injuries to be halted promptly. However, artemisinin cannot remove the infection, neither can it reduce the inflammation. Nonetheless, if artemisinin is used together with antibiotics and immunosuppressive drugs (e.g. rapamune), the treatment of rheumatoid arthritis can be significantly improved.

Anti-cancer effect of artemisinin

Cancer cells rely on nitric oxide to resist attacks by anti-cancer drugs. Through modifying the anti-cancer drugs, they decrease the effectiveness of the drugs. When artemisinin combines with the heme of nitric oxide synthase in cancer cells, it can strongly inhibit the synthesis of nitric oxide, thereby greatly increasing the sensitivity of cancer cells and bacteria to anti-cancer drugs or antibiotics.

It is currently not known whether nitric oxide, which has a protective function for the cancer cells, come from eNOS, nNOS, or iNOS. However, as the binding of artemisinin to the heme proteins is non-selective, it can be deduced that eNOS, nNOS and iNOS are all suppressed. Having lost the protective function of nitric oxide, cancer cells can be destroyed by the anti-cancer drugs.

The dosage of artemisinin is decisive in determining whether cancer cells will live or die. This is because while a high concentration of artemisinin can inhibit the synthesis of nitric oxide and cause the death of cancer cells, a low concentration of artemisinin can induce the synthesis of nitric oxide and allow cancer cells to survive.

As the anti-cancer activity of artemisinin is relatively low, it is more suitable for use to improve the effectiveness of anti-cancer drugs. If artemisinin is used together with antioxidants, its anti-cancer activity can be greatly increased. For example, the tumour inhibition rate with artemisinin alone is 47%, but when it is used with glutathione-depleting agents, glutathione peroxidase inhibitors and catalase inhibitors, the tumour inhibition rate rises to 67%.

The reason for this is that while artemisinin stimulates oxygen free radicals, it also induces the expression of antioxidases and the synthesis of antioxidants. After repeated use of artemisinin, the antioxidant capacity of cancer cells will greatly increase, thereby eliminating the oxygen free radicals and rendering artemisinin less effective or ineffective against cancer.

Anti-bacterial effect of artemisinin

Bacteria also rely on nitric oxide to block the attacks of antibiotics through modifying antibiotics and causing them to lose their effectiveness.

Gram-positive bacteria have their own nitric oxide synthase (bNOS) and can autonomously synthesise nitric oxide. Gram-negative bacteria do not possess nitric oxide synthase but rely on the reduction of nitrates and nitrites to produce nitric oxide. Although artemisinin is not able to inhibit the synthesis of nitric oxide by gram-negative bacteria (e.g. *E. coli*), it can inhibit the activity of catalase in the cells, thereby increasing the concentration of hydrogen peroxide and eventually killing *E. coli*.

Therefore, artemisinin can combine with the nitric oxide synthase in bacteria and cause it to lose its ability to synthesise nitric oxide, and it can also combine with catalase in the cells to prevent the breakdown of hydrogen peroxide. These actions will deprive the bacteria of the protection of nitric oxide, or cause them to be damaged by hydrogen peroxide, so that they can be attacked, suppressed and killed by antibiotics.

Although artemisinin has direct anti-bacterial effects, it is not very strong. Hence, artemisinin is best used to increase the sensitivity towards antibiotics, and it can effectively prevent the development of antibiotic resistance in bacteria. As artemisinin does not bind with any specific antibiotic, it is compatible with any antibiotic to raise its anti-bacterial effectiveness.

The research into the use of artemisinin against other diseases is still only at the animal trials stage and will need more time before it is available for clinical use. As more research is done, the use of artemisinin for the prevention and treatment of other diseases is likely to increase. Now that artemisinin has been awarded the Nobel Prize, there will be great enthusiasm to discover wider uses for artemisinin. Let us wait and see!

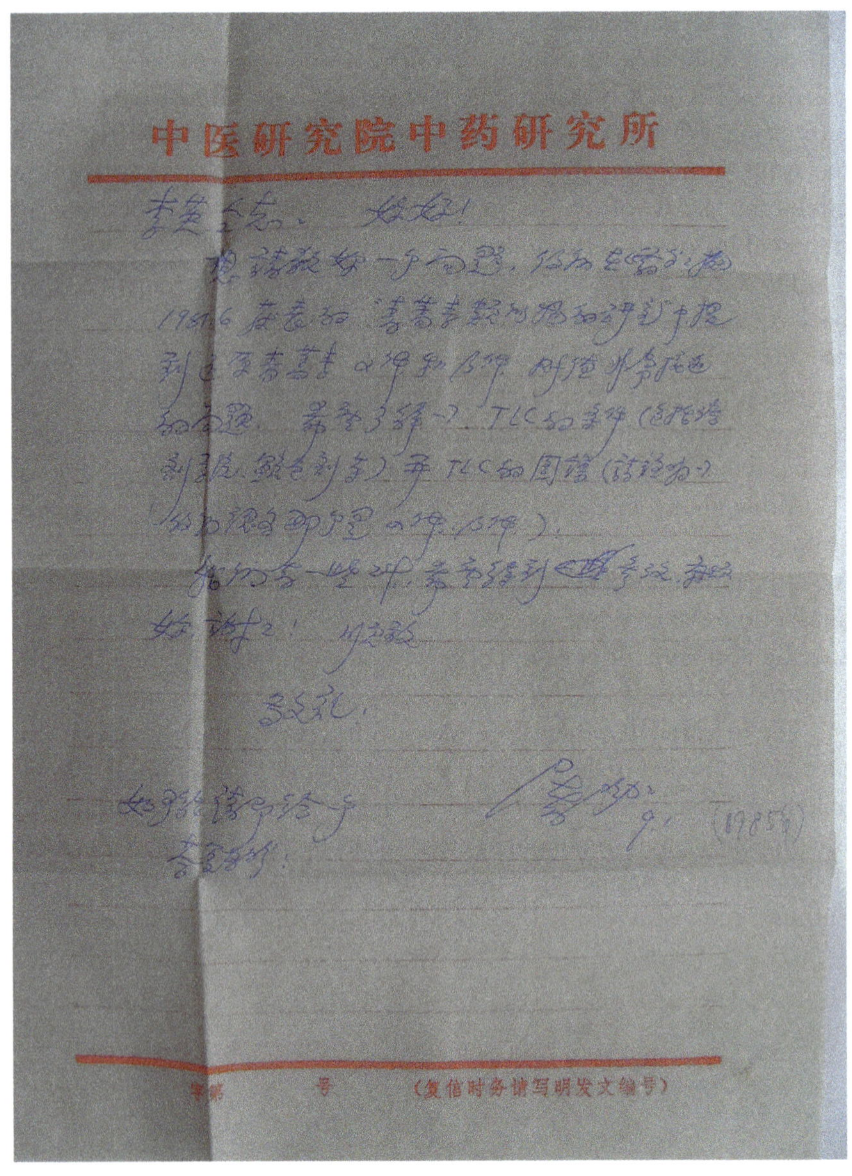

中医研究院中药研究所

字第　　号　　（复信时务请写明发文编号）

A letter from Tu Youyou to Li Ying, pharmacist and member of Project 523, on 1 September 1985

Artemisinin was included in "The 365 'Firsts' – a Pictorial Memoir of the Republic's First 50 Years"

Chapter 5
The Nobel Prize and Beyond

Artemisinin is a gift from Traditional Chinese Medicine to the people of the world. It is a contribution of great significance in preventing and treating malaria and other infectious diseases, as well as protecting the health of the people of the world.

The discovery of artemisinin is an example of a successful collective effort to explore Chinese medicine, and to win a prize based on this is an international honour for the Chinese people, Chinese science and Chinese medicine.

— Tu Youyou

The Revolution is not Yet Over

By Zhang Tiankan

The Nobel Committee for Physiology or Medicine awarded the Nobel Prize to Irish medical researcher William C. Campbell, Japanese researcher Satoshi Omura and Chinese pharmacologist Tu Youyou. The ceremony was held at the Karolinska Institute in Stockholm at 11.30am on 5 October 2015. These three researchers had developed revolutionary therapies against some of the most destructive parasitic diseases in the world.

The significance of the Nobel Prize

Tu Youyou was awarded the Nobel Prize for discovering and explaining the mechanism of *Artemisia annua* (*A. annua*) in the treatment of malaria, i.e. the effect of artemisinin. The award of the Nobel Prize to Tu Youyou made history for China and the Chinese people as this was the first time a homegrown Chinese scientist had won the Nobel Prize in the natural sciences. This has fulfilled a long-standing dream of winning a Nobel Prize.

The significance of Tu Youyou's winning of the Nobel Prize is, first and foremost, the acknowledgement that artemisinin has prevailed against malaria. Parasitic diseases have plagued mankind for thousands of years and have become a critical global issue. This is especially so as parasitic diseases affect the most impoverished populations in the world and are a heavy burden to bear while pursuing greater health and happiness. The research of the three Nobel laureates, including Tu Youyou, have revolutionised the treatment of the world's most-feared parasitic diseases, bringing hope to patients and their families.

The Nobel Prize in Physiology or Medicine is awarded for discoveries in basic medical sciences and clinical medicine. The former lays the foundation while the latter is the application. In the history of the Nobel Prize in Physiology or Medicine, more prizes have been awarded for basic medical sciences than clinical medicine. However, the 2015 Nobel Prize in Physiology or Medicine was awarded to projects involving the treatment of parasitic diseases. This shows that the Nobel Prize in Physiology or Medicine does not favour foundational knowledge over practical applications. There is, in fact, a tradition of balance between these two branches in the history of Nobel Prize.

For instance, in 1943, American Selman Waksman extracted, from a culture of *Streptomyces*, the antibiotic streptomycin which is particularly effective against *Mycobacteria tuberculosis*. This ushered in a new era in the treatment of tuberculosis and brought under control an illness that had plagued mankind for thousands of years. For that discovery, Selman Waksman was awarded the Nobel Prize for Physiology or Medicine in 1952.

Malaria is a tenacious disease. Past treatments usually made use of chloroquine or quinine, but their efficacy was declining. By the late 1960s, anti-malarial treatment methods had become ineffective while the number of malaria cases was increasing. At that time, Tu Youyou started looking for a malaria treatment among traditional Chinese herbs. Through large-scale screening of herbs and observing their effects on malaria-infected animals, Tu Youyou found that an extract of *A. annua* displayed a therapeutic effect. However, this effect was not reproducible in repeated experiments.

Tu Youyou reviewed the ancient literature again and developed an effective extraction method, becoming the first scientist to discover the extract later named artemisinin. Artemisinin showed high effectiveness against malaria in both animals and humans as it was able to kill the malaria parasite (*Plasmodium*) quickly and in the early stages.

The discovery and use of artemisinin has saved millions of lives globally. In 2001, the World Health Organization (WHO) recommended the use of artemisinin-based combination therapies (ACTs) in 76 countries severely plagued by malaria. By 2007, 69 out of those 76 countries had adopted the WHO recommendations.

Malaria, acquired immune deficiency syndrome (AIDS), and cancer are listed by the WHO as the three diseases with the highest mortality rates. The use of artemisinin has saved countless malaria patients unable to afford expensive anti-malarial drugs. In 1986, artemisinin was officially awarded a New Drug Certificate in China. In May 2004, the WHO officially listed artemisinin-based combination drugs as the preferred treatment for malaria. Since then, the "Chinese wonder drug" artemisinin has displayed its great efficacy all over the world.

According to WHO statistics, more than 3.4 billion people in the world are at risk of contracting malaria, and over 450,000 people die from malaria every year, a majority of them children. Since 2000, in malaria-plagued Africa, 24 million people in Sub-Saharan Africa have benefited from ACTs, and 1.5 million deaths from malaria have been prevented.

Due to the use of artemisinin and dihydroartemisinin, the global mortality rate from malaria decreased by 47% from 2000 to 2013. In Africa, the mortality rate decreased by 54%. As most sufferers of malaria are children below the age of five, artemisinin plays an important role in protecting the lives of children. The use of artemisinin has reduced the global mortality rate from malaria for children below the age of five by 53%, and in Africa, this rate has decreased by 58%. Nevertheless, malaria remains a major threat to mankind. In 2013 alone, 453,000 people in the world died from it.

Similarly, Omura and Campbell made great contributions toward the treatment of two other parasitic diseases: river blindness (onchocerciasis) and lymphatic filariasis. They discovered a new drug, avermectin, and its derivative, ivermectin, which have saved tens of millions of lives, especially in the impoverished regions of Sub-Saharan Africa. From 1970 to present, approximately one-third of the population living in riverside villages in West Africa could have been blinded by onchocerciasis before reaching adulthood, and about 10 million people suffer from lymphatic filariasis globally.

The contributions by Omura and Campbell led to the discovery of new and effective anti-parasitic drugs, avermectin and ivermectin. The latter is highly effective in both animals and humans against a variety of parasitic diseases. According to a WHO forecast, it is hoped that the free use of ivermectin in Africa can bring about the eradication of river blindness by around 2020. If this is realised, it could be another great achievement in medical history, similar to the victories over smallpox and poliomyelitis.

In view of the above achievements, the Nobel Assembly commented that "the global impact of their (Omura, Campbell and Tu) discoveries and the resulting benefit to mankind are immeasurable".

The revolution is not yet over

At the press conference to announce the 2015 Nobel Prize for Physiology or Medicine, an Indian reporter asked whether Tu Youyou's receipt of the Nobel Prize signified that the western medical world had changed its perception of traditional alternative medicine. With regard to this issue, a member of the Nobel Assembly, Professor Hans Forssberg, twice responded negatively. In his final reply he said, "We are not giving the prize to traditional medicine, we are giving the prize to the person who has been inspired by traditional means and developed a new drug".

The award affirms Tu Youyou's efforts in the process of drug discovery and her and other Chinese scientists' work in the discovery of artemisinin. It is not only a recognition for the work done, but also a commendation for the researchers for their dedication and sacrifices for the sake of science.

Besides looking to the past, the WHO and the Nobel Assembly recognised that it was also important to look to the future. The past rewarded in anticipation of a better and brighter future. In order to achieve this, it is imperative to press on without slackening, constantly preparing for the next challenge. This is because even artemisinin, with its high efficacy against malaria, faces the danger of losing its effectiveness. In many regions, the malaria parasite has developed resistance to artemisinin. The award to Tu Youyou and artemisinin is a call to related professionals to embark on an even greater challenge — to find a drug or treatment method that is even more effective against malaria and other parasitic diseases, in order to protect public health and lives.

Artemisinin is still being used against malaria in many regions, especially in Africa, and is notably effective in children. Without artemisinin, many African children would not survive their childhood years. The impact can be seen in that many African children have names related to artemisinin. Unfortunately, artemisinin resistance was observed as early as the 1950s and 1960s.

The discovery of artemisinin by the team of scientists led by Tu Youyou had, for a period of time, solved the problem of drug-resistant malaria. Knowing that in the constant struggle and battle by humans and drugs against parasites, the latter often evolve to overcome the drugs used against them, WHO repeatedly reminded the health departments of all countries to use artemisinin with caution, and recommended ACTs. Nevertheless, artemisinin resistance inevitably appeared.

Between 2003 and 2004, the first cases of artemisinin resistance (and ACTs resistance) appeared in the Thailand-Cambodia border regions. By 2009, ACTs were losing their effectiveness in Thailand and Cambodia. The WHO had no choice but to admit that, over the past decade, artemisinin had been losing effectiveness against malaria in Cambodia, Myanmar, Vietnam, Laos, and the Thai border regions.

This development has left many feeling helpless and regretful as artemisinin had been the most effective drug against malaria for several decades. It was also a Chinese herb that had been subjected to modern experimental science, including pharmacology, pathology, biochemistry, molecular biology and genetic studies, and had achieved global recognition and recommendations for use. In the early 21st century, WHO hoped that the resistance of malaria to ACTs was due to resistance to the companion drugs rather than to artemisinin itself.

However, it has been proven that the malaria parasite is becoming resistant to artemisinin. Extensive research has shown that the genes of *Plasmodium* have been evolving since the introduction of artemisinin, and that these changes are responsible for its resistance to the drug.

As early as 2012, a research article published in the journal *Nature* pointed out that the resistance of *Plasmodium* to artemisinin is closely related to the genetic mutation of a windmill-shaped protein known as K13. In 2013, another article published in *The New England Journal of Medicine* mentioned that the artemisinin-resistant *Plasmodium* in Southeast Asia possessed widespread K13 gene mutation. Other researchers confirmed that the K13 gene was on the thirteenth chromosome of *Plasmodium*.

Faced with artemisinin-resistant *Plasmodium* and the risks that malaria brings to health and lives, mankind must make use of greater wisdom to deal with the situation and find a solution. Especially since Tu Youyou, representing Chinese researchers, was awarded the Nobel Prize in Physiology or Medicine, the Chinese people should bear a greater responsibility to solve this problem. Other than the clinical use of a combination of drugs to reduce the risk of drug resistance, what is of foremost importance is to explore the possibility of designing compound drugs based on artemisinin or to search the treasures of traditional Chinese medicine for another anti-malarial drug so as to avoid or reduce drug resistance in *Plasmodium*. This would be a further contribution from China to the world, and possibly an opportunity for China to win another Nobel Prize (perhaps a Chemistry Prize).

Other researchers have proposed another avenue — genetic engineering. For example, the use of the newest CRISPR/Cas genetic editing system (predicted to win Nobel Prize in the future) to remove the *Plasmodium* genes responsible for drug resistance could enable artemisinin and other anti-malarial drugs to once again be effective against malaria.

Tu Youyou's award of the Nobel Prize in Physiology or Medicine this year is no doubt a joy for the Chinese, but "the revolution is not yet complete, and comrades must still work hard!"

The conflict between Chinese and Western medicine

Many Chinese regard Tu Youyou's Nobel Prize as a victory for Chinese medicine, but the Nobel Assembly repudiated this viewpoint. Professor Hans Forssberg of Sweden's Karolinska Institute, a member of the assembly, said that of the many ways to search for a new drug, mankind had long sought new remedies from different plants, and that this could stimulate a new perspective in the search for new drugs.

The answer given by the Nobel Assembly raises two questions. Indeed, the discovery of artemisinin was not directly related to Chinese medicine. However, all modern experimental medicine, including western medicine, traces its roots to the medicine of various ethnic groups. In the same way, Tu Youyou's winning of the Nobel Prize can also be regarded as being closely intertwined with Chinese medicine.

On the face of it, Tu Youyou's award is an achievement for Chinese medicine. However, it also reflects the need for modern Chinese medicine to become more aligned with current methods of experimental medicine and scientific testing. One of the tenets of modern experimental medicine is the need to satisfy Koch's postulates, i.e. that a certain hypothesis or principle has to be proven correct and be reproducible, and should be capable of being explained by modern science. In another words, there is a need to know not only how it works, but why it works.

Since ancient times, the Chinese have known that *A. annua* is effective against malaria, but little was known about its mechanism. Tu Youyou and her team extracted artemisinin with ether, which has a boiling point under 60 degrees Celsius, and performed 191 experiments before they achieved 100% effectiveness against *Plasmodium* in malaria-infected mice and monkeys in the laboratory. These experiments confirmed her hypothesis that only fresh *A. annua* extract obtained at low temperatures had a high anti-malarial effect.

Tu Youyou's award is also an affirmation of the collective achievements of all Chinese scientists, especially specialists in pharmacology, biochemistry, organic chemistry and other fields, acknowledging that scientific research requires the entire scientific community to work towards a common goal. A fuller understanding of artemisinin's mechanism was only attained because of the efforts of other scientists who continued the research.

For example, Zhou Weishan of the Shanghai Institute of Organic Chemistry of the Chinese Academy of Sciences (CAS) led a research group to analyse the structure of artemisinin and devise how to synthesise it. The research group made use of high resolution mass spectrometry to establish that the artemisinin molecule contained 15 carbon atoms, 22 hydrogen atoms and 5 oxygen atoms ($C_{15}H_{22}O_5$), belonged to the class of sesquiterpene compounds, had a sesquiterpene lactone structure with a peroxide group, and did not contain nitrogen.

Furthermore, the research group refuted the view of western researchers that anti-malarial chemical compounds that did not contain a nitrogen atom would not be effective. This conclusion was the result of modern experimental scientific research with the collaboration of scientists from different branches of science. As these early conclusions about artemisinin's chemical structure were obtained from spectrometry, it was necessary to confirm whether artemisinin truly existed by attempting to synthesise it and conducting further research into its structure. Zhou Weishan and his group later began to research the total synthesis of artemisinin.

In early 1984, after five years of painstaking research, the research group successfully achieved the total synthesis of artemisinin, and the synthesised artemisinin was exactly the same as naturally-obtained artemisinin. This was also accomplished by modern experimental science and the collective effort.

Tu Youyou's award also demonstrates that artemisinin is intricately related to Chinese medicine. As the Nobel Committee admitted, the discovery of artemisinin was inspired by Chinese medicine, albeit having undergone testing and confirmation by modern experimental medicine. This is clear from the work of Tu Youyou and the hundreds of Chinese researchers who took part in the discovery and extraction of artemisinin. For instance, as early as the second century BCE, the pre-Qin Dynasty medical book *Prescriptions for Fifty-two Diseases* already contained records of the plant *A. annua*. In 340 BC, Ge Hong from the Eastern Jin dynasty described for the first time the anti-pyretic effect of *A. annua* in his *Handbook of Prescriptions for Emergencies*. Li Shizhen also explained in his book *Compendium of Materia Medica* that *A. annua* was able to "treat chills and fevers from malaria".

Tu Youyou and her team systematically reviewed ancient medical literature and consulted experienced Chinese physicians before they shortlisted more than 640 herbs, including *A. annua*, creating a *Compilation of Anti-Malaria Herbs for Testing*. They eventually focused on *A. annua*. Therefore, it was traditional Chinese medicine which provided the clue that led to the use of modern experimental science to explain the mechanism of the new drugs artemisinin and dihydroartemisinin.

Why was Tu Youyou selected for the Lasker Award?

Long before the 2015 Nobel Prize, on 24 September 2011, Tu Youyou received the Lasker Award. This was a controversial decision.

Some felt that since Tu Youyou was neither the first to discover the anti-malarial effect of the *A. annua* extract nor the first to isolate the anti-malarial monomer, it was neither fair nor reasonable to give her all the credit. This was a valid view, and it raised the question: Why did the Lasker Awards Committee choose Tu Youyou?

This issue can be considered from two perspectives. Firstly, whether the selection process for the Lasker Award was fair and just, and secondly, whether Tu Youyou, as an individual, played a critical role in the discovery of artemisinin.

The first question is easy to answer as the selection process for the Lasker Award is similar to that for the Nobel Prize. Candidates are nominated by American and global professional bodies and there is no need to apply. This removes the need for candidates to make use of connections or canvass for support, unlike the selection process for membership of the CAS or for promotions. This alone shows that the selection of Tu Youyou for the Lasker Award was fair. In addition, 25 distinguished scientists from multiple disciplines are carefully selected to form the Lasker Jury to ensure a professional and authoritative decision. The selection process is also kept strictly confidential to ensure impartiality.

As for the second question, according to confidential materials that were later made public, although the discovery of artemisinin was a collective effort, Tu Youyou was regarded as having played the critical role. In scientific research, to play a critical role means to have come up with an idea, implemented it and proved it.

At the time, despite having shortlisted thousands of anti-malarial plants and medicines, the researchers had not found one that was ideal. The results of the experiments on *A. annua* had been unsatisfactory and the researchers had planned to give up on it. It was Tu Youyou, inspired by a line in the book *Handbook of Prescriptions for Emergencies* in the chapter "Prescriptions for the Treatment of Chills and Fever from Malaria". The line "A handful of *A. annua* immersed in two litres of water, wring out the juice and drink it all" led Tu Youyou to consider that high temperatures could affect the active ingredients in *A. annua* and thus reduce its efficacy. Tu Youyou then went on to verify this through experiments.

Tu Youyou and her team changed to using ether, which has a boiling point under 60 degrees Celsius, and after 191 experiments, they observed 100% efficacy against *Plasmodium* in malaria-infected mice and monkeys in the laboratory. So, although Tu Youyou was neither the first to discover the anti-malarial effect of the *A. annua* extract, nor was she the first to isolate the anti-malarial monomer, she was the one who established the most effective and practical method to extract artemisinin.

This is the essential point in research and can be illustrated by an interesting story. In the early twentieth century, the American company, Ford, encountered a fault in one of its electrical generators. They sought help from many technicians but none was able to locate the cause. The company had no choice but to stop production. In desperation, Ford engaged the famous electrical engineer Charles Steinmetz who, after careful examination, made a mark on the casing of the electrical generator with a piece of chalk, and instructed the technicians to open up the casing and remove 16 windings of the field coil at the location of the chalk mark. That rectified the fault. For that, Steinmetz charged a fee of US$10,000. As the highest-paid employee at Ford was only drawing a monthly wage of US$5 then, some felt that Steinmetz was overcharging. To that, Steinmetz gave a simple explanation: to make the chalk mark cost US$1; to know where to make the mark cost US$9,999.

In the same way, Tu Youyou, who was able to discover and extract artemisinin, is analogous to Steinmetz who knew where to make the mark. This was why the Lasker Jury chose her for the award.

On the other hand, while fame, seniority and experience are no doubt important in research as in other spheres of work, Tu Youyou's winning of the Lasker Award shows that the most important factor is the ability to identify and solve problems. This can be illustrated by the life of Koichi Tanaka, one of the recipients of the 2002 Nobel Prize in Chemistry. Tanaka was an ordinary engineer in Shimadzu Corporation in Kyoto, Japan. At the time he received the award, he was neither a professor nor the holder of a doctorate or a Masters degree, and could be considered to be at the lowest level of the Japanese corporate hierarchy. He also did not have any important publications to his name, and only had a few articles published in less prominent meetings and magazines. He had almost no contact with Japanese academia and would not have received any nominations from them.

It was probable that Tanaka was nominated by American or German academics, and he was in fact nominated at the last moment to replace a German researcher as the Nobel Assembly believed that the original thinking behind the method of mass spectrometric analysis of biological macromolecules originated from Tanaka. In other words, the idea was his.

Tu Youyou's situation was perhaps similar to that of Tanaka. She lacked the three important attributes of a professor: a doctorate, an overseas educational experience and membership in either the CAS or Chinese Academy of Engineering. Other than her discovery of artemisinin, everything else about her was ordinary. The awarding of the Lasker Award to Tu Youyou was similar to the awarding of the 2002 Nobel Prize in Chemistry to Tanaka.

Why was the Nobel Prize awarded to Tu Youyou alone and not her team?

After receiving the 2015 Nobel Prize in Physiology or Medicine, Tu Youyou said, "This is not an honour just for myself, but an honour for all Chinese scientists. After our research over several decades, this award is not unexpected." However, there are those who feel that it is unfair. The main question they have is why Tu Youyou alone received the award although the discovery of artemisinin was a result of a collective effort. Others who also made important contributions did not receive the award, and there were many others who made important contributions following on from her work. Examples of some of these are Luo Zeyuan, Wei Zhenxing, Li Guoqiao, Zhou Weishan and Li Ying.

Actually, this issue had been discussed since 2011 when Tu Youyou received the Lasker Award as some felt that to give credit Tu Youyou alone was unfair. As her fame grew, so did the criticism. This controversy could intensify now that Tu Youyou has been awarded the Nobel Prize.

To address this, it is important to first examine the judging criteria for the Nobel Prize as well as to understand western culture. The main judging criteria for the Nobel Prize are originality (the first to discover or invent) and significance. This arises from high value that western culture places on individual creativity and pioneering results. In other words, the judging criteria for the Nobel Prize embody western culture's emphasis on creativity.

In the discovery and development of artemisinin, Tu Youyou has no rival in terms of creativity or significance. This was clearly indicated by the Lasker Jury when they explained the reasons for their decision. The Jury felt that Tu Youyou's research had achieved three "firsts". She was the first to incorporate artemisinin into the Project 523 research, the first to extract artemisinin, and the first to perform clinical trials of artemisinin.

These reasons show that, in terms of creativity or significance, no one surpassed Tu Youyou or made contributions as great as hers.

Of course, chance played a part for the people who also made important contributions in the discovery of artemisinin but were not awarded the 2015 Nobel Prize. In the history of the Nobel Prize, there is no lack of examples of people who made important contributions but did not receive awards. It could be said that they were unlucky, or that the Nobel Committee had faced a difficult decision. Tu Youyou, together with Omura and Campbell, were considered to fall into the same category. They were awarded the Nobel Prize for important contributions in the prevention and treatment of parasitic diseases. As to why the Nobel Committee did not award the Nobel Prize in Physiology or Medicine entirely to the discovery of artemisinin, only the Committee knows.

Some Chinese argue that if the Nobel Prize was awarded for the discovery of artemisinin alone, then three Chinese could have won the Prize. However, even if that was the case, there would still have been many others who made important contributions but did not receive the award. It can only be said that the Nobel Committee was not convinced that there was anyone other than Tu Youyou who deserved the award.

A similar difficult decision was faced by the Nobel Committee when awarding the 1962 Nobel Prize in Physiology or Medicine to Maurice Wilkins, Francis Crick and James Watson for their discovery of the DNA double helix structure. Another person who should have also been awarded the Prize was Rosalind Franklin. Franklin and Wilkins were the first to shoot X-ray diffraction images of DNA which suggested that the structure of DNA was possibly a double helix. It was only later that Crick and Watson built the model of the DNA double helix structure. It can be said that the establishment of the theory of DNA's double helix structure was the outcome of the work of these four scientists, with Franklin taking the greatest credit.

It was felt that Franklin was just unlucky to have missed the award as she had died at a young age (aged 37) in 1958 from ovarian cancer. According to the will of Alfred Nobel, the Nobel Prize was to be awarded to living candidates who had "conferred the greatest benefit to mankind". Since Franklin had passed away, she could not be awarded the Prize.

At the same time, there was an unofficial rule that a Nobel Prize should not be shared by more than three persons. The discovery of the DNA double helix structure was evidently the achievement of four persons and if Franklin had been alive at the time, the judges would have faced a dilemma. Franklin's demise enabled the Nobel Committee to award the Prize to just three persons, something that satisfied all parties and avoided the Nobel Committee having to make a difficult decision.

Tu Youyou's Nobel Prize and China's failure to make nominations

Even as Tu Youyou was awarded the 2015 Nobel Prize in Physiology or Medicine, there was also a piece of news from the Nobel Committee which was not likely to please the Chinese. Professor Olga Botner, Member of the Nobel Committee for Physics, said that the Nobel Committee for Physics sent invitations to Chinese universities every year, hoping to make contact with physics professors and ask them to recommend candidates for the Nobel Prize in Physics. She said that it was a pity that there had been little response from China over the years.

If what Professor Botner said is true, a reason China not receiving a Nobel Prize for so many years could be that China had not been willing to participate in the Nobel Prize selection and hence had no voice in the Nobel Committee. This would naturally reduce the chances of winning a Nobel Prize. Even if a Chinese nominee was not chosen for the Prize eventually, it would enable others on the Nobel Committee to become familiar with Chinese research in physics and other fields, increasing the chances of winning a Nobel Prize in the future.

For many years, winning the Nobel Prize has been the dream of the Chinese. To realise this dream, ability is naturally of first importance, but other factors also come into play. Chinese research outcomes are like good wine. Although the fragrance of a good wine can lure people to a winery deep in an alley, but with the range of excellent wines available today, good wine in a deep alley will remain unknown unless it is advertised.

In reality, even if there are not many good wines, the fragrance of the wine cannot spread thousands of miles. So why did China ignore the invitations from the Nobel Committee for Physics to nominate candidates?

As the appraisal and selection process of candidates for the Nobel Prize is kept confidential, it is only possible to make some conjectures based on what is publicly known.

One possible reason could be that the Chinese physics community disdains to participate in the Nobel Prize selection and therefore ignored the invitations from the Nobel Committee for Physics. However, this is contrary to reality. In 2002, after Masatoshi Koshiba of Japan and two American scientists were awarded the Nobel Prize in Physics for their contributions to astrophysics, in particular for the detection of cosmic neutrinos, Chinese physicists were very disappointed. An article "The 2002 Nobel Prize in Physics Brushed Passed the Chinese" best describes their feelings. The article reported that in January 1978, Chinese scientist Tang Xiaowei and Koshiba had met in Germany and jointly proposed the construction of a large-scale water Cherenkov detector facility to detect proton decay. Later on, as Japan had the funds to support this project and the Chinese scientists were unsuccessful in obtaining funds, the research project was carried out by Koshiba who was later awarded the Nobel Prize. As the Chinese scientists were not able to participate in the project, they missed an opportunity to win a Nobel Prize.

Regardless of the actual details, the disappointment of the Chinese physics community was obvious. None would have rejected a Nobel Prize in Physics or any other Nobel Prize. From this, it is evident that Chinese academia did not disdain to participate in the Nobel Prize selection. Another possible cause could be that the Chinese physics community was too busy and lacked the time and energy to participate. However, this is unlikely because the Chinese yearning for a Nobel Prize is widely known. Hence, the reasons for the Chinese universities ignoring the Nobel Committee in Physics must lie elsewhere.

Another possible reason is a fear that there are no physics research outcomes in China of a standard worthy of the Nobel Prize. Each time the Nobel Committee for Physics sent the invitation to the Chinese universities, the Chinese physics community might have lacked courage to make a nomination, or been worried that the projects which they recommended would be ridiculed by others, and therefore chose to ignore the Nobel Committee in Physics.

On the other hand, it could be the Chinese physics community's lack of confidence, thinking that since they had never won a Nobel Prize personally, they were not qualified to recommend others.

There are a few other possibilities. Internal conflicts might have given rise to a situation where no one wanted anyone else to win the Prize and so nobody participated. The fear that a recommended research outcome might later prove to be fraudulent, with the risk of embarrassment, might have been a deterrent. Yet another possibility was that the Chinese physics community was occupied by fighting for projects and funds and had no time to accept the invitation of the Nobel Committee.

Of course, it would be best if none of these were the case. A lack of confidence might not necessarily be a bad thing, but it is necessary to have the courage to participate in the Nobel Committee. On this topic, British scientist Richard J. Roberts, one of the winners of the 1993 Nobel Prize in Physiology or Medicine, made a suggestion which is worth considering. He felt that one way to increase the chances of winning a Nobel Prize was to make contact with Swedish scientists. Some winners of the Nobel Prize could have been selected earlier but were delayed because they had challenged a wrong opponent or become enemies with others. One would not know if the person challenged was in the Nobel Committee or would become a member of the Nobel Committee later.

There are two points worthy of attention from this experience. One is that there is no need for China to ignore the invitations from the Nobel Committee for Physics. On the contrary, China should participate actively so that the Chinese can have a voice in the Committee. The second point follows logically, that with active participation in the Nobel Committee for Physics, the Chinese will be less likely to offend the members of the Committee, and therefore any future appraisals of Chinese physics research could be more favourable.

How great is the impact of Tu Youyou's winning of the Nobel Prize?

Being awarded the Nobel Prize is, without doubt, a great honour. However, what Tu Youyou expressed during a press interview were the hardships and difficulties of scientific research. Will this encourage students, youth and others to engage in scientific research, and change the career direction and aspirations of the younger generation?

One important issue mentioned by Tu Youyou is the monetary value of the Nobel Prize. Tu Youyou and her husband Li Tingzhao jokingly said, "The award money is not even enough to buy half a living room in a Beijing apartment! It's so little!" The total monetary value of the Nobel Prize in Physiology or Medicine in 2015 was 8 million Swedish Krona (approximately US$920,000). Tu Youyou was awarded half of it, while the other half was shared equally between William C. Campbell and Satoshi Omura.

At the US dollar to Chinese Renminbi (RMB) exchange rate of 1:6.3559 then, the US$410,000 that Tu Youyou was awarded was equivalent to about 2.6 million RMB. It was barely enough to buy half a living room of a Beijing apartment in a good location. Therefore, a research career cannot be justified by the hope of winning an award and the prize money. From another perspective, it is important to resolve the issue of researchers' livelihood before there is any possibility of research achievements. At present, those who choose to engage in research work are going against the tide.

The experience of Tu Youyou and the members of her artemisinin research team also revealed that research may involve sacrifices. At one point, Tu Youyou and her team members in Beijing found that the quality of the artemisinin was low, so they resorted to using their own bodies to test whether the leaf or stem of *A. annua* contained artemisinin.

Later, they also tested the toxicity of the drug on themselves after animal testing had shown that artemisinin had 100% efficacy against *Plasmodium*. As a result, Tu Youyou suffered liver damage and her colleagues also suffered many illnesses.

Australian Barry Marshall, one of the winners of the 2005 Nobel Prize in Physiology or Medicine, also had a similar experience. In order to prove the hypothesis that *Helicobacter pylori* (*H. pylori*) was pathogenic, he swallowed large amounts of *H. pylori* bacteria culture. Two weeks later, he developed symptoms such as gastric pains, nausea, vomiting, difficulty in eating, dizziness, cold sweats, bad breath etc. A gastroscopy found that his gastric mucosa was overgrown with slim, spiral-shaped *H. pylori* bacteria. This proved that *H. pylori* could cause gastric ulcers and Marshall was later awarded the Nobel Prize.

From these accounts, we can see that Nobel Prize winners are those who are willing to make sacrifices for science and are able to withstand hardships. If one is not willing to do so, he is clearly unsuitable to pursue scientific research.

Of course, the main aim of engaging in scientific research is not to win the Nobel Prize. In reality, of the thousands engaged in research, only a handful will be lucky enough to win the Nobel Prize. To use this year's Nobel Prize as an example, out of the 500 researchers from over 60 work units in China which played a part in developing the anti-malarial drug artemisinin, only Tu Youyou was awarded the Prize. Embarking on a career in scientific research is to accept anonymity and working on the sidelines. Like Sisyphus, one has to be prepared for a life of monotonous and difficult work.

If one does not recognise this and possess the determination to persevere, it will be impossible to pursue scientific research.

The many hardships in carrying out scientific research are reflected in Tu Youyou's experience leading up to the award. It is little wonder that many young students today are unwilling to engage in such work. Recently, a survey of high school students in China entitled "Budding Female Scientists Plan" collected responses from 1,200 students from 20 cities including Beijing, Shanghai and Guangzhou. The results showed that only 45% of high school students were willing to be scientists, while only 27% expressed a strong desire to engage in a scientific career. Only 38% of female students were willing to engage in a scientific career while the corresponding proportion for male students was slightly more than half.

Why are so few students willing to become scientists or engage in scientific research when many express a strong enthusiasm for science? The responses from the students provide some answers. The students felt that scientific research required large amounts of time and energy, and was monotonous and exhausting; that a scientific career was only for a select minority and did not relate to them; that females were a minority in the scientific community where males had the advantage; that research was not appealing; that the social status of female scientists was low; and that the recognition from society was not in proportion to the hard work.

These answers indicate that Tu Youyou's winning of the Nobel Prize would be unlikely to attract more people to engage in scientific research. However, this may be the very reason why the winning of the Nobel Prize is significant. The prize money might not allow for a comfortable life, let alone make scientists rich, but the award might inspire others to emulate scientists even if it seems difficult to attain, and may serve as a way to encourage more people to dedicate themselves to scientific research.

Naturally, the winning of the Nobel Prize confers a higher status which in turn brings with it a greater burden or responsibility which not all are able to bear. Tu Youyou said, "In comparison to winning the Prize, what gratifies me most is that the discovery of artemisinin, inspired by traditional Chinese medicine, has saved the lives of millions of malaria patients throughout the world."

If artemisinin can save the lives of millions throughout the world, and at the same time allow scientists a dignified life, for example through obtaining a patent, this may encourage more people to take up a research career.

The Importance of Scientific Thinking

By Zeng Qingping

In the 120 years since the establishment of the Nobel Prize, it was the first time a Nobel Prize was awarded to a mainland Chinese female scientist, the first time a Prize for in the natural sciences was awarded to a mainland Chinese scientist, and the first time a mainland Chinese won the Prize in Physiology or Medicine. At 5.30pm Beijing local time on 5 October 2015, Tu Youyou, the chief researcher of the Institute of Chinese Materia Medica of the China Academy of Chinese Medical Sciences, created three firsts in the history of Nobel Prize. It was then that the 1.3 billion Chinese people finally saw what was described in a Chinese poem: "only after repeated calls did she appear, with her face still half hidden behind a *pipa* (lute)".

The award for the discovery of artemisinin can be said to be unexpected, yet reasonable. To say it was reasonable was because Tu Youyou had won the Lasker-DeBakey Clinical Medical Research Award in 2011, an award which is regarded as an indicator for the Nobel Prize. It was unexpected is because Tu Youyou was not included in the Thomson Reuters list of 2015 Citation Laureates. After all, the Lasker Award for the discovery of artemisinin had been more than four years earlier, and the disappointment year after year had led people to think that it was unlikely for artemisinin to win the Nobel Prize.

Looking back through history, the artemisinin project has won many prestigious awards and its achievements have received global recognition. The success of artemisinin depended on each individual's hard work as well as collective wisdom, and the awards won comprise both individual and group awards. Other than the Nobel Prize in Physiology or Medicine that Tu Youyou shared in 2015, artemisinin has won numerous scientific awards from within and without China. In 2011, Tu Youyou was awarded the Lasker Award. In 2004, the artemisinin research group was awarded the Prince Mahidol Award, the highest award in medicine in Thailand, by the King of Thailand, Bhumibol Adulyadej. In 1996, the core members of the artemisinin research group were awarded the Hong Kong Ho Leung Ho Lee Foundation's "Seeking" Outstanding Scientific Achievement Group Award. In 1977, the artemisinin project was awarded the inaugural National Science Conference's Major Science Achievement Award.

Tu Youyou, as the outstanding representative of the Chinese artemisinin research group, is definitely worthy of the grand award. As described in the Nobel Prize presentation speech, Tu Youyou won the award because she found a new method to fight malaria — using artemisinin. Of course, anyone who is familiar with the history of the discovery of artemisinin would know that the development of artemisinin into an anti-malarial drug was not all her work alone. Her greatest contribution was to first devise an effective extraction method of artemisinin which allowed the successful analysis of its chemical structure and subsequent clinical research. Although Tu Youyou was the only one who received the award this time, the Nobel Prize was a symbol of global academia's affirmation and recognition of research in the natural sciences in China.

Tu Youyou's scientific achievements can be summarised by three "firsts". She was the first to discover the highly effective anti-malarial ether extract of artemisinin through animal and human experiments (4 October 1971). She was the first to extract artemisinin crystals from A. *annua* (8 November 1972). She was the first to obtain preliminary confirmation, through clinical trials, that the artemisinin crystals were effective on patients with malaria (September to October 1973). The 2011 Lasker Award was also given to Tu Youyou based on three "firsts". She was the first to bring artemisinin to the artemisinin research group, the first to extract artemisinin which had 100% efficacy, and the first to perform clinical trials on artemisinin.

Tu Youyou's invention of the low temperature extraction method of artemisinin was not only an innovative method, it was also an example of innovative thinking. Although the difference between high and low temperature extraction seems simple, in the days when artemisinin was unknown, this adjustment in temperature did not occur to anyone else. It was only through many experiments that the optimal temperature was established. If the temperature for the successful extraction of artemisinin had not been determined, much money, manpower and resources would have been spent on experiments with slimmer chances of success.

This shows the importance of innovative thinking in scientific research! Tu Youyou's unique way of thinking was not innate, neither was it pre-existing in her brain. Rather, it was inspired by our ancestors. In his book titled *Handbook of Prescriptions for Emergencies*, Ge Hong from the Eastern Jin dynasty wrote, "A handful of qinghao immersed in two litres of water, wring out the juice and drink it all". It was the "immersed" and "wring out" that inspired Tu Youyou to consider that artemisinin might not be destroyed by heat and therefore should be extracted at lower temperatures.

Nevertheless, as much as the experiences recorded in ancient literature are valuable, it is unwise to treat them as dogma and apply them blindly. In reality, artemisinin is insoluble in water but soluble in oil, and natural artemisinin synthesised by *A. annua* is stored in its aromatic hair glands. If extraction was done by mechanically copying the "immersed in water" method, the concentration in the artemisinin extract would be too low to be very effective.

Tu Youyou's innovation was evident on two counts. First, she changed "immersed in water" to "extracted with alcohol". As artemisinin is fat-soluble and not water-soluble, an organic solvent is more suitable for extraction. Second, she changed "ethanol extraction at high temperature" to "ether extraction at low temperature" because artemisinin loses its efficacy at high temperatures. Today, these two innovations maybe regarded as not "high-tech" at all. In fact, it would be something that would occur to most researchers in pharmaceutical chemistry. However, in the beginning when no one knew the anti-malarial component in *A. annua* or anything about its chemical structure, Tu Youyou was able to try extracting it with an organic solvent, and later with an organic solvent that had a low boiling point. This is strong proof that innovative thinking processes lie at the heart of scientific research.

The significance of artemisinin to mankind is that it has saved the lives of millions of malaria patients and has been recommended by the WHO as the core drug for anti-malarial combination therapy. The Nobel Prize is well-deserved. Artemisinin developed by the Chinese employs a completely different mechanism from that of traditional anti-malarial drugs (e.g. chloroquine) developed by other countries. It possesses superior anti-malarial characteristics. It is not only fast-acting, has high efficacy and low toxicity, and when used in combination with other drugs, it also does not induce drug resistance easily.

While artemisinin received great honour, it brought us both positive and negative revelations. From the positive perspective, the award for artemisinin fully demonstrates that the chess game model of scientific research, in which cooperation is essential, is still relevant. Combined research projects such as the nuclear and satellite programmes, insulin synthesis, and hybrid rice were all successful. Internationally, the Human Genome Project has achieved unprecedented breakthroughs through global cooperation.

At the same time, the success of the artemisinin research project illustrates that the traditional top-down (for research topics to be assigned) mechanism for scientific research can be as successful as, or even more successful, than the current bottom-up (research topics chosen by researchers) style. The motivation for the development of an anti-malarial drug arose from the urgent need on the battlefields of Vietnam when fighting the Americans, and the discovery of artemisinin came from the military-civilian cooperation research project, also known as Project 523, which was set up on 23 May 1967.

Another point to emphasise is that the success of major research projects not only depends on sufficient manpower, resources and funds, but is also dependent on the selfless dedication and noble character of a vast number of scientific workers. As the Chinese verse describes, "a high cliff is able to stand tall because it is void of worldly desires", this requires researchers to set aside the pursuit of personal glory and self-benefit, and to set their goals according to the nation's needs and service to mankind. During that time, every member of the anti-malarial research group responded to the nation's call and the needs of the revolution, and worked unceasingly in the research to discover artemisinin.

The award to artemisinin also given us cause for reflection. First, all researchers need to continue demonstrating the traditional values of self-reliance and hard work, and to strive for an entrepreneurial spirit which says "move onward when conditions allow, and even when conditions are unfavourable, create favourable conditions and still move onward". They should not constantly complain of the poor research system in the country or blame the unsatisfactory conditions of their departments. Only with dedication and selfless devotion can one achieve success and victory on the journey of scientific research. On the contrary, a person too concerned with fame will never achieve much.

As for the dispute over whether the discovery of artemisinin was an individual effort or a collective honour, the fact is that the Lasker Award and Nobel Prize have already been awarded for it. What the Nobel Prize recognised was the pioneer in the discovery process of artemisinin and not the subsequent research and clinical trials to improve artemisinin. These are the rules of the game with mainstream science, and the Chinese should learn to adapt as fast as possible. For example, the State Preeminent Science and Technology Award should be awarded to deserving individual scientists and not the group. Why do the Chinese still cast doubt on the value of an outstanding achievement which has been recognised elsewhere? Senior colleagues who participated in the artemisinin research should not be prejudiced against Tu Youyou's award but should regard it as an honour for the Chinese people. They should recognise that artemisinin has had a better reception than Chinese insulin research. The award to artemisinin should not be a cause for regret for the artemisinin researchers!

Chinese medicine was instrumental in the winning of these awards as the successful isolation of artemisinin was inspired by ancient Chinese medical literature. If there had been no recorded instructions in Ge Hong's *Handbook of Prescriptions for Emergencies* to immerse *A. annua* in water and wring out the juice to make an anti-malarial medicine, the road to the discovery of artemisinin would have been much more tortuous and difficult. It was these instructions that inspired Tu Youyou to consider that artemisinin might be destroyed by heat, and therefore, after many failed attempts to extract artemisinin at high temperatures, she decided to try using a low temperature extraction method. This not only demonstrates that Chinese medicine can be a resource for modern medicine, but also that the ancient literature of Chinese medicine is also a valuable treasure store for modern medical research. With this in mind, should we continue to pursue the modernisation of Chinese medicine? What will be the outcome of this modernisation? The achievement of artemisinin research and its numerous awards reminds us that the modernisation of Chinese medicine is of pressing urgency and we should seize the day. It should not be a question of whether to continue, but a question of how to expedite the research process.

The modernisation of Chinese medicine, with a focus on modern pharmaceutical chemistry and molecular pharmacology, is of critical importance. A common folk saying is "all plants are medicine". Every plant-based natural product is worthy of serious research and modern pharmacology has only touched on a very small portion it. Despite the importance of drug discovery and synthesis through combinatorial chemistry and high throughput screening, natural drugs developed from natural sources are even more valuable as they are highly effective yet low in toxicity. It is predicted that there will be a greater proportion of natural drugs derived from plants, animals and microorganisms available on the market in future.

How should the modernisation of Chinese medicine be carried out? Is holistic "eastern thinking" or more constituent-focused "western thinking" better? The research on artemisinin should provide a good footnote in response to this question. Some people fear that the medicinal properties of a Chinese medicine would be destroyed or greatly weakened if broken into its constituent parts, and so oppose research into the effectiveness of individual components of Chinese medicine. In fact, such concerns are groundless as the effectiveness of each component will need to be individually proven. Furthermore, it is unlikely that every component of a Chinese medicine has to be paired or matched with others in order to be effective.

There are 63 million scientific researchers in China, making up 25% of all science and technology personnel in the world. Just as Brazil has many soccer players, there is no lack of researchers in China to undertake high quality research projects which should make it possible for more Chinese Nobel Prize winners to emerge. However, it is not advisable for the country to draw up a "Nobel Prize Plan" or a "Quick Guide to Winning a Nobel Prize" as any research undertaken with the objective of winning a Nobel Prize would fall into the trap of utilitarianism. This would deviate from the original purpose of scientific research which is to understand nature and explore the unknown. Chinese American Samuel C. C. Ting, winner of the Nobel Prize in Physics, warned that scientists cannot just focus on the Nobel Prize as this would impede the achievement that could possibly lead to winning the Prize.

The results of this year's Nobel Prize for Physiology or Medicine show that the 2015 Thomson Reuters Citation Laureates prediction was completely unreliable. This was mainly because the ideals of the Nobel

Committee and other scientific research teams may differ. The Nobel Prize favours practical results, even if the research was performed many years ago. On the other hand, academic research emphasises cutting-edge research where creativity and novelty are essential. This explains why there is a time lag between the publication of research achievements and the award of a Nobel Prize. Perhaps the Thomson Reuters predictions will be true after several years. While highly influential papers will contain high quality and innovative research, papers published in academic journals with lower impact factors might not necessarily be of low quality. They may well document the preliminary stages of original or innovative research. The series of papers published in the course of the artemisinin research is a living example of that.

Chinese Medicine is a Treasure Chest

By Li Bin

Tu Youyou, professor at the Academy of Traditional Chinese Medicine (ATCM), now the China Academy of Chinese Medical Sciences (CACMS), discovered the anti-malarial drug artemisinin through searching the classic literature of traditional Chinese medicine. For that, she was recently awarded the Nobel Prize. As the news spread, the future development of traditional Chinese medicine was once again hotly debated.

How should the value of Chinese medicine be evaluated? Have we sufficiently explored the resources of Chinese medicine, especially the classic prescriptions? A reporter from the Xinhua News Agency interviewed professionals in the field of Chinese medicine and posed six questions to them.

The people interviewed were Tong Xiaolin, deputy director of the Guang'anmen Hospital, CACMS, Li Jie, head of the research department of the CACMS, Chen Baogui, ex-dean of the Tianjin Wuqing Chinese Medical Hospital and a doctor with 50 years' experience, Zhu Wanhua, head of the Chinese medicine chemistry laboratory of the Fujian Province Chinese Medical Research Institute, Zhu Benfeng, chairman of the Ding Xiang Yuan Chinese Medicine Forum, Chen Qiguang, team leader of the Chinese medicine national conditions research group, Chinese Academy of Social Sciences, and Zhang Xiaomin, assistant researcher from the same group.

Chinese medicine still has much vitality

Q: How should the value of Chinese medicine be assessed in the light of Tu Youyou's award of the Nobel Prize for the discovery of the anti-malarial drug artemisinin?

Tong: Tu Youyou's winning of the Nobel Prize will have a great impact on reclaiming this offspring of Chinese culture (Chinese medicine), recovering pride and identity in Chinese culture, and recognising that Chinese medicine has cultural significance not only to China but also to the world.

Li: In our joy, it is necessary for us to consider how we should pass on the inheritance of Chinese medicine and also innovate. This honour is a recognition as well as an encouragement. Chinese medicine is a great treasure and we should be less impulsive and carry out research with a scientific attitude. In this day and age, if everyone can reflect together and work in their specific roles, I believe Chinese medicine still has much vitality.

Zhu W: Chinese medicine is the valuable accumulation of experience over thousands of years by the Chinese people in their struggle against diseases. Classics like *The Emperor's Inner Canon* and the *Compendium of Materia Medica* contain much wisdom, such as the ability to determine the size of the uterus from the length of one's philtrum, or to determine the prognosis of liver diseases by looking at the ocular blood vessels.

The research into Chinese medicine is woefully inadequate

Q: Have Chinese medicine resources, especially the classic prescriptions, been sufficiently investigated? What room is there for further development?

Tong: Inheritance is the source of innovation. The biggest difference between Chinese and Western medical research is that the former has been tested on humans over thousands of years. This is unmatched in global medicine. The key to Tu Youyou's breakthrough was found in a classical Chinese medical text. Others in the Chinese medical profession must investigate more of our ancestors' wisdom for the benefit of modern medicine. To ignore or disparage the past is to betray the essence of Chinese traditional culture.

Zhu B: Completely insufficient, this is only the tip of the iceberg.

Zhang: The research into both classic or secret folk prescriptions is definitely inadequate. The medicines prepared by individual hospitals according to the classic prescriptions are a treasure. We need a system to regulate and protect these resources so that they can play a bigger role in the treatment of patients, in intellectual property rights and in the pharmaceutical market. We need to break away from the reliance on the modern medical system, and revolutionise the production of traditional medicine.

Chinese medical professionals should develop confidence from their own practice

Q: Can we have confidence in Chinese medicine? How can this confidence be encouraged?

Tong: Of course we should. Efficacy has always been the aim of Chinese medicine professionals. Professor Tu Youyou's greatest achievement was the saving of millions of malaria patients with artemisinin. The Nobel Prize is her reward from science.

Zhu W: The ancient saying goes, "If you can't be a good minister, then be a good doctor". Chinese medicine professionals need, first and foremost, to have a sense of mission and responsibility, to bear the hardships of studying, poverty, and loneliness, and even more importantly, to have team spirit. "Tranquility breeds wisdom, and wisdom breeds intelligence". The temptations in the world today are too strong and too many, and there are few who are tranquil. Chinese medicine professionals should develop confidence from their own practice.

Zhang: Faced with global medical difficulties, we should have confidence in Chinese medicine and make good use of its unique resources. We should establish a legal system that promotes the survival and development of Chinese medicine and protects the intellectual property rights in Chinese medicines. This would help solve global medical challenges in a way that is characteristic of China.

The westernisation of herbal medicines: a path to new drugs

Q: Can Tu Youyou's award be said to be a victory for Chinese medicine? What are your views on the current problems of extracting the active ingredients in Chinese medicine?

Tong: To develop Chinese medicine, there is the need to make use of the achievements of modern science and medicine. Tu Youyou's success was dependent on the techniques of modern science and medicine. It is a classic example of combining traditional and modern technology. Arrogance and complacency will inhibit the development of Chinese medicine.

Zhu W: Tu Youyou's award demonstrates that the "westernisation" of herbal medicine is a way of discovering new drugs, but it is just one way and should not replace all other research models. Chinese medical prescriptions comprise different herbs which each play a different role according to Chinese medicine principles. For example, Japan views the prescriptions in the *Treatise on Cold Damage Disorders* as treasures and produce them on a large scale in factories without the need to further test them as new drugs.

Zhang: Modern methods of extracting active ingredients can be employed, but we must not deny the theories, practices and achievements of traditional Chinese medicine. The crisis we face is that many people do not realise the importance of protecting the intellectual property rights in traditional Chinese medicines. We need to use the law to safeguard our rights of autonomy, leadership and intellectual property in our resources.

Zhu B: Tu Youyou regarded Chinese medicine as a repository of natural medical resources, and succeeded in extracting an effective monomer to create a chemical drug. There is a need for others to continue this type of research. However, the most pressing need today is to encourage high quality Chinese medical research which makes use of the philosophy and methods of Chinese medicine to study the new diseases and problems of modern society.

The potential of Chinese medicine

Q: The Nobel Assembly has claimed that the award to Tu Youyou was to commend the discovery of a new drug which was inspired by Chinese medicine. Will this set off a wave of research on phytochemical drugs by the large foreign pharmaceutical companies?

Tong: Faced with the complex diseases of human society today, for example geriatric, chronic and metabolic diseases, Chinese medicine has enormous potential. The treatment of complex diseases requires the combination of multiple ways of thinking and different techniques. Chinese medicine's holistic and personalised approach, coupled with modern medicine's systemic biology, precision medicine and individualised treatments give rise to the possibility of great innovations in both models of thought and research skills.

Zhu B: I think it will. In 1998, Pfizer had already solicited a prescription from us.

Chen: Foreign drug companies set up research centres in China long ago. Isn't it obvious that they are emulating the route of artemisinin?

Chinese medicine can be effective against intractable modern diseases

Q: How can Tu Youyou's award be an inspiration for the development of Chinese medicine? Do you have any suggestions?

Tong: Why is Chinese medicine still able to flourish today despite the strength of western medicine? The reason is simple. It is because the human body is extremely complex and the mechanisms of many diseases are still not completely understood. Especially in modern society where the chief illnesses are now chronic diseases, geriatric diseases, metabolic diseases and mental illnesses, how can they be controlled and treated holistically? In both theory and practice, modern medicine has been shown to be insufficient. On the other hand, the essence of Chinese medicine is systemic, holistic, homeopathic, using external treatment for internal adjustment, and diagnosis and treatment are highly individualised. This systemic approach can indeed be effective against some intractable modern diseases. This is the only reason that Chinese medicine is still flourishing today. China's medical reforms and medical system cannot be separated from Chinese medicine. The greatest inspiration that Tu Youyou's award has brought is the recognition that Chinese medicine belongs to the Chinese people and also to the world.

Chen: Chinese and western medicine each have their characteristics and strengths. To combine the use of both for the health of the people is the right and obvious way to go.

Artemisinin and Malaria

By Zhang Tiankan

On 12 September 2011, the winners of the Lasker Awards were announced and Tu Youyou of the CACMS was awarded the Clinical Medical Research Award for "the discovery of artemisinin, a drug therapy for malaria that has saved millions of lives across the globe, especially in the developing world". That was the highest international award that the Chinese biomedical field had ever received. Some people felt that the winning of the Lasker Award brought the Nobel Prize a step closer.

Malaria and mankind

Malaria is endemic in tropical and sub-tropical climates including China. A person with malaria will show cycles of alternating high fever and chills, hence its common name "swinging pendulum". The ancient Chinese named malaria "poisonous miasma" as they thought it was the floating mists that poisoned the sufferers.

In 1880, French doctor Charles Louis Alphonse Laveran who was working in Algeria found the presence of a unicellular parasite in the bodies of malaria patients and confirmed that it was the cause of the disease. This pathogen was the malaria parasite. For that discovery, Laveran was awarded the 1907 Nobel Prize in Physiology or Medicine. In 1897, British doctor Ronald Ross proved that malaria was passed to humans through mosquitoes, and for that he was awarded the 1902 Nobel Prize in Physiology or Medicine.

These two discoveries laid the foundation of man's fight against malaria. The malaria parasite is a cross-host protozoon. It carries out asexual reproduction in a vertebrate host, and undergoes maturation and carries out sexual reproduction in insects. Four species of the malaria parasite have been discovered to date, of which the most common and most typical is *Plasmodium falciparum* (*P. falciparum*). Malaria is spread by female *Anopheles* mosquitoes, while the male mosquitoes feed mainly on plant sap and are not involved in the transmission of malaria.

Five to ten minutes after a person is stung by a female *Anopheles* mosquitoe, sporozoites invade the human liver. By doing so, they avoid being attacked by the human immune system and make use of nutrients in the liver cells to differentiate and reproduce in large numbers. About a week later, the merozoites burst from the liver cells and are released into the blood stream. They immediately invade the red blood cells to once again avoid the immune system. They feed on haemoglobin and continue to replicate. After approximately two days, the merozoites cause the infected red blood cells to rupture and they go on to invade more red blood cells. In this way, two-thirds of the red blood cells in the infected person may be invaded by the parasite in a short time. The cyclical replication of the malaria parasite in the blood is what causes the infected host to experience alternating high fever and chills.

Some merozoites in the red blood cells develop into sexual forms of different sizes, male and female gametocytes, which will be taken up by the mosquito that stings the infected person. They survive and mature in the gut of the mosquito where they combine and reproduce. Two weeks later, the newly formed sporozoites enter the salivary glands of the mosquito and when the mosquito stings another person, it transmits the sporozoites to a new host. This is the infection chain of malaria transmission. Humans and mosquitoes are the two indispensable hosts at different stages of the malaria parasites' life cycle.

Infected persons can start to exhibit symptoms of malaria nine to fourteen days after being stung by a mosquito carrying sporozoites. Other than alternating chills and fever, the host will also experience flu-like symptoms such as headache and nausea. If prompt treatment is not given, the infection can progress quickly and endanger life. This is due to the destruction of the red blood cells resulting in severe anaemia or the destruction of important organs, for example caused by the blockage of capillaries which supply blood to the brain.

Inspired by the use of cinchona bark by the Native American tribes to treat malaria, French researchers Pierre Joseph Pelletier and Joseph Bienaime Caventou extracted the anti-malarial drug quinine from cinchona bark in 1826. This was the first western drug for the treatment of malaria. As the malaria parasite developed resistance to quinine, other new drugs like chloroquine and pyrimethamine were developed.

The discovery and extraction of Artemisinin

Nevertheless, as the malaria parasite successively developed resistance to chloroquine and pyrimethamine, it became necessary to develop another drug more effective against malaria. Artemisinin was the drug chosen and is the most effective anti-malarial drug today. The Chinese have known about *A. annua* for a long time. As early as the second century BC, *Prescriptions for Fifty-two Diseases*, a medical book from the early Qin period, already had records of the plant *A. annua*. In 340 BC, Ge Hong from the Eastern Jin dynasty described, for the first time, the anti-pyretic effect of *A. annua* in a Chinese medicine prescription book titled *Handbook of Prescriptions for Emergencies*. Li Shizhen also explained in his book *Compendium of Materia Medica* that *A. annua* could treat the chills and fevers from malaria. *A. annua*, also known among the common people as *chouhao* or *kuhao*, is a member of the plant family *Asteraceae*. In the book *Classic of Poetry*, the plant "*hao*" mentioned in the line "The deer bleat '*youyou*' while they feed on wild *hao*" is *A. annua*.

As the malaria parasite had become resistant to the anti-malarial drug quinine, China launched a massive project in 1967 that brought together the science and technology capabilities of the entire nation to develop a new anti-malarial drug. The project was launched on 23 May, and the project was subsequently named Project 523. 500 researchers from over 60 departments participated in this project. On 21 January 1969, Tu Youyou, as the leader of the research team from the ATCM, joined Project 523.

Prior to this, other researchers in China had already screened and selected more than 40,000 possible anti-malarial compounds and Chinese herbs, but none of these proved satisfactory. Hence, Tu Youyou and her team started to systematically review the ancient medical literature and consult experienced Chinese physicians, eventually compiling a list of more than 640 herbs, including *A. annua*, into a book titled *Compilation of Anti-Malaria Herbs for Testing*.

However, in the first round of drug screening and experiments, the extract from *A. annua* only had a 68% efficacy of against malaria, results that even pepper surpassed. At the same time, from the information compiled by the other research departments and submitted to the Project 523 Office, the conclusion was that anti-malarial effect of *A. annua* was not very high. In the second round of drug screening and experiments, the efficacy of *A. annua* dropped to only 12%. For quite a long time, *A. annua* was ignored.

The discovery of *A. annua* actually required perceptiveness. Tu Youyou once again reviewed the ancient literature and found a line in the book *Handbook of Prescriptions for Emergencies*, in the chapter "Prescriptions for the Treatment of Chills and Fever from All Malaria", stating "A handful of *A. annua* immersed in two litres of water, wring out the juice and drink it all". Could it be that only the fresh juice of *A. annua* had a strong anti-malarial effect? Furthermore, this record differed from the traditional method of brewing and consuming Chinese medicine. As boiling exposed Chinese herbs to high temperatures, was it high temperatures that destroyed the anti-malarial effects of the *A. annua* juice?

With these questions in mind, Tu Youyou and the researchers from her team began their experiments. They used ether, with a boiling point below 60 degrees Celsius, to perform the extraction from *A. annua*. On 4 October 1971, after 191 experiments, Tu Youyou observed 100% efficacy against *Plasmodium* in malaria-infected mice and monkeys in the laboratory. This indicated that their scientific hypothesis was correct, that only fresh *A. annua* juice extracted at low temperatures had a high anti-malarial effect.

Further verification by modern experimental science

The fresh *A. annua* juice extracted was named artemisinin. Proving that the artemisinin extracted at low temperatures had a strong anti-malarial effect was only an important first step in artemisinin research. More research was needed to explain why artemisinin had such a strong anti-malarial effect. For instance, it was necessary to determine the molecular structure of artemisinin, its molecular weight, whether it could be synthesised, its mechanism for killing *Plasmodium* etc. Using modern experimental science, other Chinese researchers continued to work on these puzzles.

In early 1973, the Beijing Institute of Materia Medica received artemisinin crystals so that the organic chemists there would be able to determine artemisinin's chemical structure and its anti-malarial mechanism. This task was later passed on to Zhou Weishan of the Shanghai Institute of Organic Chemistry of the CAS who led his research group in carrying out the structural analysis and total synthesis of artemisinin.

Zhou Weishan and his group made use of scientific techniques such as high resolution mass spectrometry. Through repeated investigations they were able to confirm that the structure of the artemisinin molecule contained 15 carbon atoms, 22 hydrogen atoms and 5 oxygen atoms ($C_{15}H_{22}O_5$), belonged to the class of sesquiterpene compounds, had a sesquiterpene lactone structure with a peroxide group, and did not contain nitrogen. This refuted the belief of western researchers that anti-malarial chemical compounds that did not contain nitrogen would not be effective. The structure analysis of artemisinin ended in 1976 and the paper "The structure and reaction of artemisinin" was published in *Acta Chimica Sinica*, volume 37, 1979.

In 1979, Zhou Weishan and his group began to research the total synthesis of artemisinin. As previous conclusions on its chemical structure had been obtained from spectrometry, total synthesis was a test to confirm whether artemisinin truly existed. This research was important to further authenticate its structure.

In early 1984, after five years of painstaking research, the research group achieved the total synthesis of artemisinin. This synthesised artemisinin was exactly the same as natural artemisinin. Soon after, the research paper "Structure and synthesis of artemisinin and related compounds" was published in *Acta Chimica Sinica,* volume 42, 1984. In 1977, the artemisinin project won the Major Science Achievement Award at the National Science Conference. In 1987, the total synthesis achievement was awarded the second prize in the National Natural Science Awards.

Anti-malarial mechanisms

Before the discovery of artemisinin, drugs such as quinine, chloroquine and pyrimethamine were already in use, each with a characteristic anti-malarial mechanism. Quinine was effective against all *Plasmodium* trophozoites in their erythrocytic stage, so it was able to control the clinical symptoms of malaria. The mechanism of action of chloroquine was more complicated. Clinical pharmacological research showed that after the use of chloroquine, the content of the drug in the lysosomes of *Plasmodium* was more than a thousand-fold higher than in the host. From this, it was believed that *Plasmodium* had a special mechanism which concentrated chloroquine within the parasite. Moreover, chloroquine was able to bind itself to the DNA double helix of *Plasmodium* to form a DNA-chloroquine complex which affected the cells' DNA replication and RNA transcription, and also broke up the RNA. In this way, chloroquine inhibited the division and replication of *Plasmodium*. Also, chloroquine, a weak base, after entering the body of the parasite, caused the pH of the cell solution to increase. This formed an unfavourable environment for the proteolytic enzymes and decreased *Plasmodium's* ability to divide and break down haemoglobin. The resultant lack of essential amino acids in turn disrupted the replication of *Plasmodium*.

Pyrimethamine was able to inhibit *P. falciparum* and tertian malaria during the primary exoerythrocytic stage. It was also able to prevent the proliferation of the *Plasmodium* sporozoites in mosquitoes, thereby controlling the spread of malaria. Although it was slow to take effect, it also inhibited the immature *Plasmodium* schizonts in their erythrocytic stage to control the symptoms of quinine-resistant *P. falciparum*. It also inhibited the dihydrofolate reductase of *Plasmodium* to block the synthesis of nucleic acids. Pyrimethamine, commonly used with dihydrofolate synthase inhibitor sulfonamides or sulfones to increase the efficacy, was used to treat chloroquine-resistant *P. falciparum*.

The anti-malarial mechanism of artemisinin is distinctly different from that of the earlier anti-malarial drugs. Although the mechanism of action of artemisinin is not fully clear, preliminary findings give some indication of the anti-malarial action of artemisinin.

The characteristic feature of artemisinin is its fast inhibition of the maturation of *Plasmodium*. *In vitro* experiments have shown that artemisinin is able to inhibit the growth of asexual *P. falciparum* and can directly kill the parasites. Furthermore, the effect of artemisinin is 1.13–1.16 times stronger than chloroquine. Experiments on drugs derived from artemisinin found that, of the three drugs artemisinin, artemether and sodium artesunate, sodium artesunate is the most effective. It is 16 times more effective than chloroquine and 14.3 times more effective than artemisinin. The anti-malarial effectiveness of artemisinin and artemether is similar to that of chloroquine.

Artemisinin works primarily on the membrane system of *Plasmodium*. First, it affects the food vacuole, superficial membrane and mitochondria of *Plasmodium* cells. It then disrupts the nuclear membrane, endoplasmic reticulum and the chromosomal bodies in the nucleus. By disrupting the superficial membrane and mitochondrial function of *Plasmodium*, it can prevent the digestive enzymes of *Plasmodium* from breaking down the haemoglobin of its host into amino acids which serve as its food. When *Plasmodium* is unable to get food, it quickly becomes starved and autophagosomes are rapidly formed. These are discharged continuously from *Plasmodium*, which loses a large amount of cytoplasm and eventually leads to disintegration and death. The intake of tritium-labelled isoleucine by *in vitro* cultivation of *P. falciparum* has also demonstrated that the anti-malarial mechanism of artemisinin could be due to the inhibition of protein synthesis in *Plasmodium*.

The mechanism of sodium artesunate, after entering the human body, is to cause a reduction of artemisinin which inhibits the functions of the limiting membrane of food vacuoles and the cytochrome oxidase of the mitochondrial membranes. By doing so, it disrupts *Plasmodium's* supply of nutrients from red blood cells. Therefore, the anti-malarial mechanism of artemisinin is distinct from that of quinine and chloroquine which work by disrupting the replication of *Plasmodium* through inhibiting its DNA, and also distinct from pyrimethamine and the sulfonamides, which work by disrupting folic acid metabolism to kill *Plasmodium*. Artemisinin works by inhibiting the functions of the limiting membranes of food vacuoles and the cytochrome oxidase of the mitochondrial membranes, thereby directly killing *Plasmodium*.

Integrated treatment methods for malaria yield the best results

Despite having strong anti-malarial effects, artemisinin has its limitations. For example, although artemisinin can kill *Plasmodium* in the erythrocytic stage, it is ineffective in the exoerythrocytic and pre-erythrocytic stages. Artemisinin can also trigger drug resistance, although not as quickly as chloroquine.

The occurrence of chloroquine drug resistance in *Plasmodium* could be related to the increase in discharge and higher metabolism of the drug in the body. On the other hand, pyrimethamine resistance in *Plasmodium* is due to the rapid increase in dihydrofolate reductase which causes its effectiveness to be greatly reduced. In the light of *Plasmodium* developing resistance to these drugs, the WHO has suggested that all countries adopt Artemisinin-based Combined Therapies (ACTs) by using quinine with artemisinin, or using an artemisinin compound drug.

Although the efficacy of ACTs is still above 90%, the WHO has confirmed that artemisinin-resistant *Plasmodium* has appeared in Cambodia and Thai border regions in recent years. Therefore, the

WHO has proposed that every country should act quickly to contain artemisinin resistance. On 12 January 2011 in Geneva, the WHO issued an action plan entitled Global Plan for Artemisinin Resistance Containment and requested that every country carry out five actions:

First, to contain the spread of drug-resistant *Plasmodium*. Second, to strengthen the supervision and detection of artemisinin resistance. Third, to standardise the criteria for use of ACTs in clinical practice. Fourth, to increase research on artemisinin resistance. The global plan specifically pointed out that funds were needed for research into artemisinin resistance with the aim of developing a faster and more effective way to detect drug-resistant *Plasmodium*, and to discover a new anti-malarial drug to eventually replace artemisinin as the foundation drug in the combination therapies. Fifth, to mobilise resources and motivate action.

It can be observed that because the mechanism of artemisinin resistance is not clear, the WHO, while considering the use of combination drugs, is also seeking a new drug that can eventually replace artemisinin. This implies that artemisinin would not be the only drug and method to fight malaria. In fact, the WHO has always emphasised prevention as well as integrated treatment methods in fighting malaria. Statistics from the WHO indicate that from 2000 to 2010, the global anti-malarial operation saved the lives of 730,000 people, of which three-quarters were during the period since 2006. This was because indoor residual spraying of DDT, insecticidal nets (mosquito nets impregnated with insecticides such as deltamethrin) and ACTs have become increasingly widely used in the past five years.

Prevention is far more economical and effective than treatment. The indoor residual spraying of DDT is currently one of the most effective tools against malaria. Every family need only spray US$5 worth of the chemical. Even the long-lasting insecticidal nets cost only US$5 each and there is no need for re-soaking within five years. Malaria has the greatest impact on the poor, with 80% of malaria cases occurring in Sub-Saharan Africa which is within the poorest 20% of the world. The use of insecticidal nets and indoor residual spraying of DDT are the most effective malarial prevention methods for the poor. Hence, drug treatment, including the use of artemisinin, is just one measure among many that are used together. Without the use of preventive measure like the indoor residual spraying and the insecticidal nets, it will be impossible to guard against malaria no matter how effective drugs like artemisinin can be.

Therefore, the best method against malaria is integrated prevention and treatment.

Certificate of Tu Youyou's National Pioneering Workers Title Award

Medal and certificate of Tu Youyou's New Century Heroine Inventor Award

Appendices

As a scientist, it is a very great honour to receive the Nobel Prize. The success of the artemisinin research project was the result of collective efforts over the years, and so the award for artemisinin is a collective honour for the scientists in China.

This award is also an expression of the international scientific community's great interest in scientific research in Chinese medicine, and is a gateway into that community. It brings me much joy, and gives China and Chinese scientists great cause for pride.

The award of this Nobel Prize confirms that Chinese medicine is a valuable resource, but it requires thoughtful consideration to explore and improve it.

— Tu Youyou

Appendix 1: Timeline of Project 523 (1964 to 1981)[1]

Time	Events
1964	Chairman Mao had a meeting with a Vietnamese communist official who mentioned the serious malaria epidemic in Vietnam. He requested help to deal with this epidemic. Chairman Mao replied: "Solving your problem would be solving our problem as well". The General Logistics Department (GLD) issued an order to the Academy of Military Medical Sciences (AMMS) and the Second Military Medical University (SMMU) to develop a long-acting anti-malarial drug. The two work units were to carry out this project concurrently.
1964 to 1967	The anti-malaria research was carried out in the AMMS, SMMU and the Institutes of Military Medical Sciences of the Guangzhou, Kunming and Nanjing military districts.
1966	From May to August 1966, AMMS deployed a large number of personnel to Vietnam to survey the overall health, incidence of disease outbreaks, and the prevention and treatment of disease among the Chinese troops stationed in Vietnam with a focus on the incidence, prevention and treatment of malaria. In June, considering that the research capability of the military academy alone was insufficient, the GLD began to coordinate a major national collaboration involving multiple organisations. A great deal of preparation work was carried out before the national collaboration meeting planned for 1967.
1967	On 4 May, the Science and Technology Commission (STC) notified the relevant work units about the Anti-Malaria Drug Research Work Meeting. On 18 May, the national leadership team for anti-malaria research held a meeting in Beijing.

[1] Li Runhong, Master's thesis, Peking University, 2011.

	From 23 to 30 May, the STC and the GLD jointly convened the first Anti-Malaria Drug Research Work Meeting. The meeting discussed and confirmed a three-year research plan. A total of 88 representatives from 37 participating work units attended the meeting. For confidentiality, the meeting date was abbreviated and the research plan was given the codename "Project 523". On 16 June, the STC and the GLD jointly issued the summary of the Anti-Malaria Drug Research Work Meeting and the Anti-Malaria Drug Research Work Plan.
1968	From 19 to 21 February, the STC, GLD, Ministry of Health (MOH), Ministry of Chemical Industry (MCI), Chinese Academy of Sciences (CAS) and relevant work units convened the 2nd Anti-Malaria Drug Research Work Meeting in Hangzhou. The meeting set out regulations for tasks and responsibilities of the leaders, division of labour between departments and the anti-malaria research collaboration. There were no major changes to the three-year plan drawn up during the first meeting in 1967. However, there was a much more detailed description of the tasks of each leadership group and formal rules as to confidentiality were set out in writing. On 29 May, the STC and the GLD jointly issued the summary of the 2nd Anti-Malaria Drug Research Work Meeting. In May, the Beijing 523 Leadership Office issued a document detailing the arrangements and regulations following from the issues discussed at the 2nd Anti-Malaria Drug Research Work Meeting. On 10 June, the new "Beijing 523 Leadership Group Office" seal was activated. After the discussion between the leadership groups, the original "Anti-Malaria Drug Research Leadership Group Office" seal would be changed to "Beijing 523 Leadership Group Office".
1969	In January, the Academy of Traditional Chinese Medicine (ATCM) joined Project 523. On 15 January, the STC and the GLD submitted an Anti-Malaria Research Report and Request to the Prime Minister and the Central Military Commission (CMC). The report suggested holding a forum in either Beijing or Guangzhou for the personnel in charge from the Revolutionary Committees of each province, city, and region, general logistics departments of each military district, and other work units.

Appendix 1 (*Continued*)

Time	Events
	On 8 February, Prime Minister Zhou signed a special telegraph. The content of the telegraph: *Notice regarding the convening of the Anti-Malaria Drug Research Forum:* *"With the approval from our great Chairman Mao, the Anti-Malaria Drug Research Forum shall convene in Guangzhou. The STC and GLD will compile a list of personnel who should participate and other related issues. The specific time and venue will be made known at a later date."*
	In March, a Project 523 forum was held in Guangzhou.
	In October, a Project 523 forum was held in Beijing.
1970	On 16 January, the Beijing 523 Leadership Group held a meeting to hear from the Project 523 offices regarding the progress of their work, and to receive reports from the four specialist teams: the drug synthesis and screening team, the traditional Chinese medicine team, the mosquito repellants team and field prevention and treatment team.
	In October, the neutral extract from the Chinese herb *A. annua* exhibited 100% efficacy in malaria-infected mice and monkeys.
1971	On 16 March, the Military Control Commission of the MOH, the MCI, CAS and the GLD submitted an Anti-Malaria Research Report and Request to the State Council and the CMC. The report suggested making adjustments to the Leadership Group, appointing the MOH to be the team leader and the GLD to be the assistant team leader, and retaining the office at the AMMS.
	On 15 April 1971, the State Council and the CMC passed the (71) State Document No. 29 to approve the request.

From 21 May to 1 June, the National Anti-Malaria Research Meeting convened in Guangzhou. During the forum, the six original organisations of the 523 Leadership Group which consisted of STC (leader), GLD (assistant leader), Commission for Science, Technology and Industry for National Defense (COSTIND), Ministry of Health, MCI and CAS were changed to Ministry of Health (leader), Ministry of Health of the GLD (assistant leader), MCI and CAS. The four organisations were to take the lead while the office remained at the AMMS. In addition, the meeting drafted a five-year plan for national anti-malaria research from 1971 to 1975. The corresponding research plan and manpower needs were adjusted accordingly. Other than other than making changes to the original research units, new work units were invited to participate in the research to meet the new research requirements. These units were the National Vaccine and Serum Institute, Beijing Medical University and the Beijing Institute of Pharmaceutical Industry. After the SMMU troops moved to Xi'an, Xi'an Pharmaceutical Factory began collaborating in the research. This was a significant meeting as it enabled Project 523 to continue, eventually leading to the rediscovery of the anti-malarial properties of *Artemisia Annua (A. annua)*.

From 2 to 28 July, the Guangdong Regional 523 Office collaborated with the National 523 Leadership Office and the personnel in charge of the relevant work units in the field research team to check on the teams at the Hainan site. The main focus was to strengthen the leading group and to solve the on-site collaboration issues.

On 19 July, the seal of the National Anti-Malaria Drug Research Leadership Group Office was put into use.

From 11 to 14 August, the Guangdong 523 Regional Office convened a Chinese medicine Anti-Malaria Forum. The participating work units present which had been conducting Hainan field research were six medical and military groups from Beijing, Shanghai, Guangxi and Guangdong, as well as personnel from six health units from the Hainan Military District and province. The meeting summarised the progress of the work and enabled participants to exchange experiences. A few Chinese herbs which showed higher efficacy were mentioned, such as *Artabotrys hexapetalus*, the root of *Brucea javanica* and *Hydrangea macrophylla*.

Appendix 1 (*Continued*)

Time	Events
	From 27 October to 2 November, a meeting to summarise the Project 523 field work in Hainan was held in Haikou. A total of 14 teams comprising 103 people from Beijing, Shanghai, Sichuan, Guangxi and Guangdong had participated in the field research work in Hainan. Seven teams and leadership groups, as well as 64 office personnel participated in the meeting. The meeting pointed out the main problems of the field work then: the work done by the office was not thorough and detailed enough, they did not remain long in the field and did not provide comprehensive guidance; there was limited exchange of experiences and information between the teams carrying out field work; the preparation carried out by each team was insufficient and teams withdrew too early from the field; some of the workers made *ad hoc* changes; the research into anti-malarial Chinese medicine was progressing too slowly; and there was a lack of confidence and little progress in the research on acupuncture and new medical treatment methods against malaria.
1972	In March, the Anti-Malaria Drug Research Specialist Teams Meeting (Drug synthesis and screening team and Traditional Chinese medicine team) convened in Nanjing. Substance of the reports at the meeting: the chemical structure of *Artabotrys hexapetalus* was being analysed after exhibiting promising anti-malarial results during clinical trials; preliminary separation of the effective chemical monomer of *Agrimonia pilosa* Ledeb. had begun and its chemical structure would be analysed if it showed effective results in clinical trials; effective extracts had been obtained from the following herbs: *A. annua*, *Strigose Hydrangea* leaves, *Hydrangea macrophylla*, *Nandina domestica*, *Aristolochia yunnanensis* Franch., and *Euphorbia helioscopia* and were undergoing laboratory and clinical research to improve their efficacy. The meeting recommended three actions: to analyse the chemical structure of *Artabotrys hexapetalus* and to continue research on its chemical synthesis; to determine the chemical structure of *Agrimonia pilosa* Ledeb. if the clinical efficacy of the effective monomer was confirmed; to hasten the extraction and investigate the clinical efficacy of the effective compound or monomer of *A. annua*, *Ailanthus altissima* and other selected herbs.

In March, the specialist team on mosquito repellants held a meeting in Shanghai. The meeting commented on the need to further integrate western and Chinese methods, to encourage the widespread usage of plant-based mosquito repellants, and to promote and improve the work.

From 24 August to early October, researchers from the ATCM conducted a clinical trial of the neutral ethyl ether extracts of *A. annua* in the low malaria-endemic region of Changjiang, Hainan. The clinical trial consisted of 11 cases of tertian malaria among persons from outside the area, 9 cases of *P. falciparum* malaria, and 1 case of mixed infection. Chloroquine was used to treat 3 cases of *P. falciparum* malaria. The tertian malaria cases were regarded as controls.

On 10 November, the National Anti-Malaria Drug Research Leadership Group Meeting convened in Beijing. Xie Hua from the Military Control Commission of the MOH, Wu Heng from the CAS, Chen Zixing from the MCI, Yang Dingcheng from the GLD Health Department, Long Dashi from the GLD Technology Department, and relevant personnel in charge attended the meeting. Xie Hua presided at the meeting. Other than summarising the work by the leadership groups, the meeting further discussed the problems faced in the anti-malaria research. The meeting suggested that the anti-malaria research project be incorporated into the national research plan. Hence, the relevant departments were requested to have this research included in the department and system research plans. The leadership and organisations were also strengthened, and regulations to govern the publication of research results were made.

From 20 to 30 November, the forum for the directors of regional 523 Offices convened in Beijing. The personnel in charge and specialist representatives from work units under the Beijing regional office also participated. Xie Hua from the National Anti-Malaria Drug Research Leadership Group and the Military Control Commission of the MOH, Chen Zixing from the MCI, Wang Mengzhi from the CAS, Wang Erzhong from the GLD Health Department, Liu Yinsheng from the GLD Technology Department and other leaders attended the meeting. Representing the leadership group, Xie Hua presided at the meeting.

By the end of the year, the ATCM had separated and extracted multiple monomers. One of the extracted monomers that exhibited anti-malarial properties was named “Artemisinin II”.

Appendix 1 (*Continued*)

Time	Events
1973	From 4 to 12 January, a forum on malaria immunity was held in Shanghai. A total of 63 representatives from Beijing, Shanghai, Sichuan, Guangdong, Jiangsu, Yunnan, Guangxi, and Guizhou attended the forum.
	From 15 to 22 January, a research forum regarding the treatment of *P. falciparum* malaria convened in Guangzhou. A total of 53 representatives from eight provinces and cities of Guangdong, Yunnan, Shanghai, Beijing, Nanjing, Guangxi, Sichuan and Guizhou attended the forum. In addition, the personnel in charge of the National Anti-Malaria Drug Research Leadership Group Office and the Guangdong Regional 523 Leadership Group attended the forum.
	On 15 February, the MOH, MCI, COSTIND, and the GLD submitted an Anti-Malaria Research Report and Request to the Prime Minister. It reported on the progress of work since the passing of (71) State Document No. 29 by the State Council and the CMC, and the five-year plan. It also reported on the testing of the prescription given by the French doctor, General Riche, as instructed by the Prime Minister. The conclusions were: to satisfy the requirements of China and Vietnam's anti-malaria needs, three types of anti-malaria tablets should be made widely available in endemic regions; to incorporate the anti-malaria research into the national research programme; to request for permission to convene another Anti-Malaria Drug Research Forum; and to adjust the plans for the final three years of the five-year research plan.
	On 20 February, a research work meeting on insecticides and insecticidal equipment was held in Shenyang.
	In April, researcher Luo Zeyuan from the Yunnan Institute of Materia Medica brought some *Artemisia absinthium* back to the research laboratory for extraction. An effective anti-malarial monomer was obtained through ethyl extraction. It was temporarily named "*Kuhao* Crystal III" and was later named "*huanghaosu*".

	The Shanghai Anti-Malaria Drug Research Forum convened from 28 May to 7 June 1973. This meeting was attended by the personnel in charge from the MOH, the science and education group of the State Council, MCI, ATCM and GLD, the leaders and specialists from various provinces, cities, autonomous regions, military districts, and work units. Representatives of the Chinese Communist Party (CPC) Schistosomiasis Leadership Team offices and the Departments of Commerce of the 13 southern provinces, cities and autonomous regions also attended. A total of 86 people attended the meeting. From September to October 1973, the ATCM carried out clinical trials in Changjiang, Hainan using the artemisinin II extract. Out of the 8 cases of tertian malaria and *P. falciparum* malaria they observed, 3 cases were among the non-local population. The dosage per capsule was 3–3.5g, and the average time taken for the fever to subside was 30 hours. Reviews were done after 3 weeks: 2 of them achieved full recovery, treatment was effective in 1 case (*Plasmodium* parasites resurfaced 13 days later). For the 5 *P. falciparum* cases among the non-local population, treatment was effective for 1 case (more than 70,000/mm^3 of *Plasmodium* parasites, dosage was 4.5g in tablet form, the fever subsided in 37 hours, presence of parasites in blood film turned negative in 65 hours, *Plasmodium* parasites resurfaced 6 days later); due to premature heart contractions, 2 cases stopped the medication early (this was the first time for 1 of the cases, with 30,000/mm^3 of *Plasmodium* parasites, a dosage of 3g, the fever subsided in 32 hours, *Plasmodium* parasites and fever resurfaced 1 day after stopping the medication); treatment was ineffective in the other 2 cases. In November, the Shandong Academy of Chinese Medicine managed to extract the effective monomer from *Artemisia annua* L. obtained from Shandong province. The monomer was named "*huanghuahaosu*".
1974	From 1 to 17 October, a forum for personnel in charge of the regional 523 offices was held in Beijing. The personnel in charge of the units responsible for the project from Beijing, Xi'an and Shenyang also attended the forum. The meeting summed up the research work done by each specialist group and made further plans for future research work.

Appendix 1 (*Continued*)

Time	Events
	In February, the ATCM sent Ni Muyun to the Shanghai Institute of Organic Chemistry (SIOC), bringing some research documents and artemisinin samples. This was a collaboration with the SIOC to analyse the chemical structure of artemisinin.
	From 28 February to 1 March, the researchers of the four units conducting research into *A. annua* from Beijing, Shandong and Yunnan came together for an *A. Annua* Research Forum with the 523 Office and the ATCM leaders.
	From 15 to 25 April, a specialist meeting on the drug synthesis was held in Shangqiu city, Henan. Specialists from Beijing, Shanghai, Nanjing, Guangdong, Yunnan, Sichuan, Guangxi, Shenyang and Xi'an attended the meeting. Representatives from the provinces and military districts of Henan, Shandong, Anhui and Wuhan also attended the meeting. A total of 65 representatives were present.
	During the first half of May, the Shandong *Artemisia Annua* L. Research Collaboration Team carried out clinical trials at the Zhuzhuang Brigade, Guandong Commune, Juye County, Shandong. *Huanghuasu* was used on 10 cases of tertian malaria. The drug was given in capsule form, each capsule containing 0.1g of the crystal. The cases were split into 2 groups, each comprising 5 people. One group consisted of 3 adults and 2 children between the ages of 10 and 12 years. Dosage used was 0.2g for the adults and 0.1g for the children once per day, to be consumed for 3 consecutive days. The other group consisted of 5 adults, dosage was 0.4g once per day, to be consumed for 3 consecutive days. The conclusion from this first clinical trial performed by the Shandong Province *A. annua* L. Research Collaboration Team was that *huanghuasu* was a fast-acting anti-malarial drug which could be used for emergency treatment. There were no adverse effects observed during the treatment process. However, it was not a complete cure as the relapse rate was fairly high. As it would not be easy to control the rate of relapse with *huanghaosu* alone, the conclusion was that it would be necessary to use *huanghaosu* concurrently with other anti-malarial drugs.

	From September to November, the Yunnan team treated 3 cases: 1 case of *P. falciparum* malaria and 2 cases of tertian malaria; Guangdong Academy of Chinese Medicine 523 team treated a total of 18 cases: 14 cases of *P. falciparum* malaria (including 3 dangerously-ill cases) and 4 cases of tertian malaria. After the clinical trial, the Guangdong Academy of Chinese Medicine 523 team concluded that *huanghaosu* was a fast-acting anti-malarial drug. An initial dosage of 0.3–0.5g was sufficient to quickly control the maturation of the *Plasmodium* parasites. The reason for the early return of the symptoms and parasites might be due to a rapid excretion rate (or the rapid rate the drug was converted to other compounds within the body). As the drug concentration in the blood was not maintained for a sufficiently long period, it was unable to completely eliminate the *Plasmodium* parasites. It was also mentioned for the first time that *huanghaosu* had the characteristics of a highly effective and fast-acting drug that could be used to save patients who had dangerous forms of malaria. In November, a meeting to assess and evaluate the research on mosquito repellants was held in Shanghai and technical evaluation reports were produced.
1975	In February, a forum for personnel in charge of the 523 Office was held at Beijing Rainbow Hotel. The meeting mainly summarised and assessed the Project 523 work in 1974 and discussed the coordination and implementation of the 1975 work plan. From 14 to 24 April 1975, the Project 523 Traditional Chinese Medicine Forum convened in Chengdu. A total of 62 representatives from the Project 523 traditional Chinese medicine research teams of Beijing, Shanghai, Jiangsu, Guangdong, Guangxi, Sichuan, Yunnan and Shandong attended the forum. Representatives from relevant work units from Henan, Hunan and Hubei, and several senior Chinese physicians and rural medical personnel were also present. Each research unit reported on the progress of their research. There was a special focus on the traditional Chinese medicine research team from the Guangdong Academy of Chinese Medicine. Over a period of eight years, they had penetrated deep into malaria-endemic rural villages, accumulated experience in treating cerebral malaria and obtained commendable results. The meeting also commented that some research units focused on laboratory research, and had a tendency to work behind closed doors.

Appendix 1 (*Continued*)

Time	Events
	In November, an *A. Annua* Research Forum convened in Beijing. The members of the 523 Leadership Group and researchers from Beijing, Shandong, Yunnan, Guangdong, Sichuan, Jiangsu, Hubei and Shanghai attended the meeting. They reported their research results and progress, and drew up the 1976 *A. annua* research plan. The plan included research on sources of *A. annua* plants, simpler dosage forms of *A. annua*, the effective component of *A. annua* in malaria treatment and the problem of relapse, the analysis of the chemical structure of artemisinin II, and other chemical components of *A. annua*. On 30 November, Liang Li and the others from Institute of Biophysics of the CAS managed to determine the chemical structure of artemisinin and its corresponding structure using a diffraction method. From 10 to 20 December, the Project 523 Drug Synthesis Evaluation Meeting was convened in Shanghai. The meeting was attended by representatives from Beijing, Shanghai, Guangdong, Yunnan, Sichuan, Shandong, Henan, Hubei, Zhejiang, Shanxi, Shenyang and Huainan who had participated in seven projects involving drug research, production and clinical trials. A total of 37 people were present. The leaders from the MOH MCI, CAS, and the Department of Health of the GLD also invited the relevant leadership departments and personnel in charge from the Shanghai Regional 523 Leadership Group, the National Vaccine and Serum Institute and Shanghai Institute for Food and Drug Control to attend the meeting.
1976	On 20 February, the ATCM made a report to the MOH Leadership Party Group. “A New Type of Sesquiterpene — Artemisinin” was published in the *Chinese Science Bulletin,* issue 3, 1977. The report was reproduced by the US publication *Chemical Abstracts*. This was the first publication about artemisinin.

1977 From 18 to 28 February, the ATCM and Shandong Provincial Academy of Integrative Medicine jointly launched a discussion forum on the Chemical constituents of artemisinin in Jinan, Shandong. A total of 20 researchers from the following work units participated in the meeting: ATCM, Shanghai Institute of Materia Medica, Guangdong Academy of Chinese Medicine, South China Institute of Botany, Sichuan Institute of Chinese Medicine, Yunnan Institute of Materia Medica, Institute of Botany of Jiangsu Province, Health Bureau of Gaoyou County of Jiangsu Province, Henan Provincial Institute of Food and Drug Control, Henan Pharmaceutical Factory, Institute of Parasitic Diseases of the Hubei Academy of Medical Sciences, Hubei Jianmin Pharmaceutical Factory, Shandong Provincial Academy of Integrative Medicine, Shandong Institute for Drug Control, and Shandong Institute of Scientific and Technical Information. During the meeting, the work units introduced more than 10 different methods to determine the constituents of artemisinin, but every method had inherent problems. During the meeting, it was suggested that each work unit use the colorimetry and capacity methods as a reference and continue their research to further improve their methods.

From 21 to 30 March, the Project 523 Work Forum convened in Beijing. Representatives from the provinces, cities, and regions of Beijing, Shanghai, Guangdong, Guangxi, Yunnan, Sichuan, Jiangsu, Shandong, and Henan attended the meeting. Representatives from the military districts of Kunming, Guangzhou, Nanjing, Chengdu and Guangxi, Shanghai Garrison, Chinese Academy of Medical Sciences (CAMS), ATCM, AMMS, the four military universities, Northeast Pharmaceutical Factory, and 71 personnel in charge of Project 523 offices also attended the meeting. During the opening and closing sessions, Jiang Yizhen from the MOH, deputy minister Tao Tao from the MCI, Tian Ye from the CAS and deputy minister Zhang Xiang from the GLD made speeches. Other than hearing reports of the research progress and exchanging experiences, the meeting also drew up the anti-malaria drug research work plan for 1977 to 1980.

From 22 to 29 April, the National 523 Office convened the Specialist Group Meeting on Integrative Medicine for Malaria Treatment in Nanning, Guangxi. The meeting summarised and evaluated the research into *A. annua* since the meeting in Chengdu in 1975. The meeting specifically detailed the tasks to be completed before the final evaluation of the results obtained. This purpose of this meeting was to prepare for the final evaluation of artemisinin research.

Appendix 1 (*Continued*)

Time	Events
	In June, the Specialist Group Meeting on Drug Synthesis was held in Shanghai to evaluate a drug to treat cerebral malaria, Naonuejia, and another malaria drug, hydroxy piperazine.
	From 14 to 21 December, the Specialist Group Meeting on mosquito repellants convened in Guangzhou. A total of 63 representatives from research units, health organisations and universities from Beijing, Shanghai, Guangdong, Guangxi, Yunnan, Sichuan, Jiangsu, Shandong, Henan, Hubei, and Zhejiang, and the military districts of Guangzhou, Nanjing and Shenyang attended the meeting.
1978	The No. 3 anti-malaria drug, piperazine and pyronaridine tetraphosphate received the 1978 National Science Conference major achievement award. Hydroxy piperazine tablets and hydroxy piperazine phosphate received the advanced technology achievement award at the National Science Conference. Pyronaridine tetraphosphate and changrolin received the National Science Conference award.
	On the basis of anomalous X-ray diffraction scattering data, researchers from the Institute of Biophysics of the CAS meticulously analysed and succeeded in confirming the molecular structure of artemisinin.
	From 9 to 16 May, the Specialist Forum on Malaria Immunity convened in Chengdu. The meeting summarised the research experiences and progress over the past few years. The research plans for 1978 to 1985 were discussed and agreed upon.
	In June, for the very first time, *Guangming Daily* published information on the new anti-malaria drug — artemisinin.

	From 23 to 29 November, a meeting was held to appraise the research results using artemisinin (and *huanghuahaosu*) in Yangzhou, Jiangsu. The meeting was hosted by the National Anti-Malaria Drug Research Leadership Group. Leaders and representatives from the MOH, STC and GLD attended the meeting. The leaders from provinces, cities, regions, military districts, regional 523 offices involved in the research task also attended the meeting. The leaders and main technical personnel from nine provinces and cities, and the troops that participated in *A. annua* and artemisinin research, medical and epidemic prevention units, medical academies, drug manufacturing factories, and the main research units and collaborating research units were also present. Invitations were also extended to representatives from the Chinese Medical Association, Pharmacopoeia Commission of the MOH, National Institutes for Food and Drug Control and the *New Medical Journal*.
1979	On 4 September 1979, the State Administration of Medicine document (79) Drug No. 387 proposed the handover of drug synthesis research to the AMMS. From 1980 onwards, MCI was no longer in charge of the military medical research. Additionally, the proposal mentioned the decreasing number of tasks handled by Project 523 and acknowledged that the project was a collaborative effort between both military and civilian organisations. From 1980 onwards, all civilian medical research plans would no longer involve the military medical organisations. In September, the new anti-malarial drug, artemisinin, received the National Invention Award (second class). On 15 October, the 4th edition of the *People's Daily* reported that the manufacture of the new anti-malarial drug, artemisinin, invented by the CACMS, Shandong Academy of Chinese Medicine, Yunnan Institute of Materia Medica, Institute of Biophysics of the CAS, SIOC, and Guangdong Academy of Chinese Medicine has been awarded the National Invention Award (second class).
1980	On 13 June, Huang Shuze, Deputy Minister of the MOH, hosted the National 523 Leadership Group Meeting in Beijing. The relevant personnel in charge from the STC, State Administration of Medicine, GLD, MOH, and AMMS attended this meeting. The meeting affirmed the 523 research work and the results obtained over the past 13 years. The meeting also set out regulations to govern the work after the national and regional Anti-Malaria Drug Research leadership groups and organisations had been dissolved.

Appendix 1 (*Continued*)

Time	Events
	On 25 August 1980, the four leading departments, the MOH, STC, State Administration of Medicine and GLD jointly requested instructions from the State Council and CMC to include the anti-malaria drug research project in the provincial and city research plans of the relevant ministries, and to revoke the national collaboration work between organisations. The document was signed by Huang Shuze, Zhao Dongwan, Huang Kaiyun and He Biao.
	On 27 August 1980, the four leadership departments, the MOH, STC, State Administration of Medicine and GLD jointly requested that the State Council and the CMC revoke the national anti-malaria research collaboration and to incorporate the anti-malaria drug research project into the plans of relevant organisations in the provinces, cities and states. The document was signed by Huang Shuze, Zhao Dongwan, Huang Kaiyun and He Biao, and reviewed by Qian Xinzhong. It was then approved by Deputy Prime Minister Chen Muhua, with the agreement of Wan Li and Fang Yi.
	On 25 November, the Science and Technology Division of the MOH issued a notice regarding the allocation of the prize money for the invention of artemisinin. The notice dealt with issues pertaining to the allocation of the prize money and certification. The 5,000 RMB prize money was to be allocated as follows: 2,200 RMB to the ATCM, 1,000 RMB to the Shandong Academy of Chinese Medicine, 1,000 RMB to the Yunnan Institute of Materia Medica, 400 RMB to the Guangdong Academy of Chinese Medicine and 200 RMB each to the SIOC and the Institute of Biophysics of the CAS. The notice also instructed that these institutions to pass 10–15% of the prize money to the work units they had collaborated with. These were mainly work units in Hainan which had collaborated with the CACMS, work units in Henan which had collaborated with Shandong Academy of Chinese Medicine, and other work units in Yunnan which had collaborated with the Yunnan Institute of Materia Medica.

1981	From 3 to 6 March, the Anti-Malaria Drug Research Meeting for the leadership group and the personnel in charge was held in Beijing. The meeting in 1981 mainly executed the changes to the 523 organisation structure. However, the anti-malaria project remained an important one and was included in the regular research plan of relevant departments from the states, provinces, cities, autonomous regions and military. In view of the adjustments made to the Project 523 organisation structure, the MOH set up a Malaria Committee within its Medical Sciences Committee. The military also decided that the GLD would organise and plan for an anti-malaria team within the Epidemic Specialist Team in May 1981. The MOH, STC, State Administration of Medicine and GLD jointly issued awards to the participating units and to individuals in the leadership groups. A total of 134 work units (collective) received awards. 17 units were in the science and technology system, 55 units were in the medical and health system, 27 units were in the pharmaceutical and chemical system, 26 units were in the military system, and 9 units were in the light manufacturing, higher education and other systems. A total of 85 individuals also received awards; they came from seven regions: Beijing, Guangdong, Guangxi, Nanjing, Shanghai, Sichuan and Yunnan. On 11 May, representing the Anti-Malaria Drug Research Leadership Team, the four leading departments jointly issued a final document — the National Anti-Malaria Research Forum Summary. Besides a summary of the meeting, the notice also outlined the plan for the winding up of Project 523. The regional leadership group administration departments would determine the handling and transfer of the 523 Office documents, technical files, funds and supplies, and also make arrangements with the original work units of the full-time Project 523 personnel who had been away for a long period of time.

Appendix 2: Milestones in Artemisinin Research (1967 to 2003)[2]

Time	Event
1967	On 23 May, the Anti-Malaria Drug Research Leadership Group and the Project 523 administrative body was set up by the government to solve the problem of drug-resistant malaria. In the course of the project, a total of over 60 research units and 500 researchers nationwide participated in Project 523. They came from research institutes, universities, pharmaceutical factories and hospitals under different systems and ministries. The research areas of Project 523 included developing new types of mosquito repellants, drugs to prevent, treat and eradicate malaria, acupuncture treatment and mosquito-exterminating equipment. There were two approaches in the plan to develop new drugs. One was to focus on the modification of current anti-malarial drugs and synthesise the anti-malarial drugs used overseas. The other was to find and test new anti-malarial drugs from Chinese herbal medicine.
June 1967 to 1971	The anti-malarial ingredient, beta-febrifugine, extracted from the Chinese herb *Dichroa febrifuga*, exhibited a stronger efficacy in killing the *Plasmodium* parasite than quinine. However, beta-febrifugine caused severe vomiting in patients. Right from the beginning, one of the main aims of Project 523 was to modify the chemical structure of beta-febrifugine. Numerous beta-febrifugine derivatives and similar compounds were synthesised and screened. Subsequently, the beta-febrifugine derivative (no. 7002) and changrolin (*no*. 56) underwent clinical trials, but they were not ideal anti-malarial drugs. Concurrently, a large number of Chinese herbs underwent extraction and screening.
1971	The ethyl extract of *Artemisia annua* L. (*huanghuahao*) was proven through animal experiments to have anti-malarial properties.

[2] Abstracted from Tu Youyou *Artemisia annua and artemisinin drugs*, Chemical Industry Press, 2009.

1972	In March, a meeting of the drug synthesis and traditional Chinese medicine specialist groups was held in Nanjing. During the meeting, it was reported that more than 10 types of anti-malarial compounds or monomers had been extracted from *Agrimonia pilosa* Ledeb., *Artabotrys hexapetalus*, *Artemisia annua* L., *Polyalthia nemoralis*, *Hypericum japonicum* Thunb., *Brucea javanica* and *Nandina domestica*. From August to October, the ethyl extract of *Artemisia annua* L. was used in a small-scale clinical trial in Hainan. Its high efficacy, fast-acting properties and the ability to treat cerebral malaria made it the main focus in the national anti-malarial drug research.
1973	The crude formulation of the ethyl ether extract was tested on tertian malaria cases in Shandong. Its anti-malarial effect was stronger than chloroquine and side effects were mild. Through the separation and purification of the effective anti-malarial monomer of *Artemisia annua* L., a type of white needle-like crystal was obtained. After that, research on its chemical structure began.
1974	Large amounts of the effective anti-malarial monomer were extracted from *Artemisia annua* L. Following pharmacological and toxicity experiments, the monomer was used in large-scale clinical trials in Yunnan and Shandong for the treatment of *P. falciparum* and tertian malaria. The monomer was fast-acting and exhibited high efficacy with low toxicity. However, while it was capable of treating chloroquine-resistant *P. falciparum* malaria, the rate of relapse was quite high.

Appendix 2 (*Continued*)

Time	Event
1975	In April, the Project 523 Traditional Chinese Medicine Forum was held in Chengdu, Sichuan. The national collaboration was organised to launch the research on the effective monomer of *Artemisia annua* L. During the meeting, there was much debate as to what to name the effective monomer. Suggestions included *huanghuahaosu*, *huanghaosu*, or *qinghaosu* (it was later formally named *qinghaosu* or artemisinin). The peroxide compound structure of *yingzhaosu*-A was also reported. This work provided an inspiration for the work on the structural analysis artemisinin. A few days later, artemisinin was proven to contain a peroxide group. According to spectral data and chemical reaction experiments, it was deduced that artemisinin was a type of sesquiterpene lactone with a peroxide group. The by-product, dihydroartemisinin, obtained when artemisinin was reduced with sodium borohydride, laid the foundation for subsequent synthesis and preparation of artemisinin derivatives. After the Chengdu meeting, the search for sources of wild *Artemisia annua* L. was carried out nationwide. The technique and method of using gasoline solvent for extraction matured gradually. By the end of the year, the molecular structure and the relative structure of artemisinin were determined using X-ray crystallography. The following year, the absolute structure of artemisinin was determined via anomalous X-ray crystallography. The chemical structure of artemisinin was completely different from other known anti-malarial drugs. In December, the the Project 523 Drug Synthesis Evaluation Meeting convened in Shanghai. The meeting evaluated the results of the clinical trials of the derivatives of beta-febrifugine (no. 7002), *changrolin* (no. 56), pyronaridine (no. 7351), *naonuejia*, and nitroquine (no. CI679). The results obtained were satisfactory, but their low solubility made it difficult to be used as injectable formulae for the treatment of patients with severe malaria. Also, as the rate of relapse was high, improvements were needed.

1976	In February, the work into the modification of artemisinin's structure began. In December, the quality standards for artemisinin were formulated.
1977	In April, the Specialist Group Meeting on Integrative Medicine for Malaria Treatment was held in Nanning, Guangxi. In June, the Specialist Group Meeting on Drug Synthesis was held in Shanghai. The Quantitative Structure — Activity Relationship (QSAR) model of artemisinin and ideas for the design of derivatives were reported during the meetings, together with the results of the first batch of more than 20 artemisinin derivatives tested on malaria-infected mice. Their effectiveness against malaria was several times that of artemisinin. An article on artemisinin was published in the *Chinese Science Bulletin* under the name of Artemisinin Structure Research Group, making public the chemical and relative structure of artemisinin.
1978	The artemether oil injection was tested in a clinical trial in Hainan, while the sodium artesunate injection was tested in Guangxi. Both injections achieved better anti-malarial efficacy than *A. annua* and were also faster-acting and more reliable in the treatment of *P. falciparum* malaria patients. In November, a meeting was held in Yangzhou, Jiangsu to assess the research on artemisinin. The meeting summed up the conclusions as to the sources, chemistry, pharmacology, formulations, clinical experiments, manufacturing techniques and quality control standards of artemisinin.
1979	The invention of the new anti-malarial drug, artemisinin, was awarded a National Invention Award (second class). An article on the anti-malarial properties of artemisinin was published under the name of the Artemisinin Structure Research Group in the National Medical Journal (English edition). This was the first time the data regarding the pharmacological and clinical research on artemisinin's anti-malarial properties was made known to the public.

Appendix 2 (*Continued*)

Time	Event
1981	The Meeting to Evaluate the Anti-Malaria Drug Artemether was held in Shanghai. In October, the World Health Organization (WHO) held and hosted a meeting on Artemisinin and its Derivatives in Beijing.
1985	On 1 July, the Ministry of Health issued the New Drug Evaluation and Registration System.
1987	The oil-based artemether injection and the sodium artesunate injection were approved for production.
1988	The Sichuan Wuling Shan Pharmaceutical Factory started constructing a factory for artemisinin production. The construction was completed in 1990, and it became the world's largest artemisinin production plant.
1989	In April, the WHO hosted the meeting of the Scientific Working Group on the Chemotherapy of Malaria in Beijing.
1992	The dihydroartemisinin tablet and artemether-lumefantrine compound were approved for production.
1997	The oil-based artemether injection was listed in the 9th edition of the WHO's *Model List of Essential Medicines*.
1998–2004	The artemisinin-naphthoquine and dihydroartemisinin-piperaquine compounds were approved for production.
2001	In November, the WHO and MOH hosted the Meeting on Anti-Malarial Drug Development in Shanghai.
2002	The artesunate tablet was listed in the 11th edition of the WHO's *Model List of Essential Medicines*.
2003	The artemether compound was listed in the 12th edition of the WHO's *Model List of Essential Medicines*

Appendix 3: Artemisinin — Patents and Collaboration

By Zeng Qingping

The regret of failing to patent artemisinin

Why did no-one patent the wonderful invention, artemisinin? This is one question frequently asked. Looking back, the reason why artemisinin was not patented was neither "foreign agents" nor "Chinese traitors". It was a result of mistakes in decision-making within the organisations and the divulging of information by the researchers themselves.

During that era, the main impetus was to exert collective effort to bring honour to the country; there was no sense of individualism or personal gain. During the mid-1970s in China, after the Project 523 Meeting, the anti-malarial and chemical properties of artemisinin were already clear. However, for the sake of confidentiality, no articles were published. Even if any article had been published, the regulations of the organisations would have required that it be published under the name of a collective group.

In 1972, at the 8th International Symposium on the Chemistry of Natural Products held in New Delhi, India, a botanical chemist from Yugoslavia presented a research report which unexpectedly mentioned that a new type of sesquiterpene lactone had been extracted from *Artemisia annua (A. Annua)*. The molecular formula and molecular weight of the sesquiterpene lactone obtained was exactly the same as the results for artemisinin obtained by China! However, it was not until 1976 that the Chinese heard that the Yugoslav scientists had managed to separate similar products from *Artemisia* plants. The Chinese thought that the sesquiterpene lactone obtained by the Yugoslavs from *A. annua* was a compound identical to artemisinin, so they felt impelled to publish their research.

In order to publish their findings before the Yugoslavs and prove that artemisinin was a Chinese invention, the Health department of the China Academy of Chinese Medical Sciences (CACMS) requested and received approval from the Ministry of Health (MOH) to publish a paper under the name of the Collaboration Research Group for Artemisinin making public the chemical and relative structure of artemisininin. The paper appeared in the *Chinese Science Bulletin*, volume 22, issue 3, 1977. The title of the article was "A New Sesquiterpene Lactone — Artemisinin". The publication of this major paper made public the structure of artemisinin, and China lost the intellectual property rights to this unique and precious chemical compound.

The reason for this mistake in decision-making was the result of the lack of adequate knowledge in China about international patent and intellectual property rights. The leaders did not realise that it was possible to apply for an international patent for artemisinin, and no one was sent to visit Yugoslavia to investigate their research before making the decision. According to the Yugoslav scientist who had researched the chemical composition of A. *Annua* and who visited China in 1986, "Even if we had managed to determine the exact chemical structure, we would not have developed it into an anti-malarial drug." This was because Yugoslavia did not have any experience in using A. *annua* for malaria treatment. What was more ironical was that although they had determined the correct molecular formula and weight of the chemical compound, they made an error in the chemical structure. They had mistakenly identified this compound as dihydroartemisinin ozonide. Hence, the Chinese scientists' fears were groundless. It was a false alarm!

Fortunately, the first paper published by China did not disclose the anti-malarial properties of artemisinin. Furthermore, the paper "The Crystal Structure and Absolute Configuration of Artemisinin" published under the name of the Collaboration Research Group for Artemisinin and the Institute of Biophysics of the Chinese Academy of Sciences (CAS) in *Sciencia Sinica*, issue 11, 1980, likewise did not mention these anti-malarial properties. However, after this, many researchers in China went on to publish papers within China and elsewhere, revealing the anti-malarial properties of artemisinin to the world. This naturally resulted in a link being drawn between the chemical structure of artemisinin and its anti-malarial properties.

The publication of these articles by the researchers leaked some key information. Although the structure of artemisinin had been made known and was no longer eligible for patent protection, the anti-malarial properties of artemisinin (e.g. usage, dosage, formulations) could still have been patented. This would have mitigated the loss from the first mistake. However, once this information was leaked, it was public knowledge. The researchers continued to publish clinical information and trial results within China and internationally, and by doing so, selflessly made a gift of our national treasure to the west.

Nevertheless, although this might have seemed unfortunate at first, it did no damage to the country in the long run. The loss of the artemisinin patent may even be a blessing in disguise. Malaria is prevalent in many poorer regions, and to profit from the poor in these undeveloped countries would be inhumane. Since this anti-malarial drug is not protected by a patent, it can be sold cheaply or given freely to the malaria patients in poorer regions. This is both humanitarian and charitable.

The difficulties of collaboration

After China made public the molecular structure of artemisinin, the results of the clinical trials of artemisinin were published in domestic and foreign journals. This attracted the attention of the WHO which had been concerned over the uncontainable spread of chloroquine-resistant malaria. The general director of the WHO sent a letter to the MOH asking to convene a symposium on artemisinin and its derivatives in China as soon as possible.

During the Cultural Revolution, China had been in a state of isolation with no idea what was happening outside the country. Foreign countries were also unfamiliar with the situation in China. Realising that such an exchange would be mutually beneficial for both parties, the Chinese government promptly agreed. Thus, a Symposium on Artemisinin and its Derivatives was hosted by the WHO Scientific Working Group on the Chemotherapy of Malaria was convened in Beijing. A total of seven research reports were presented at the meeting, all by representatives from China. One of the reports, "Chemical Studies on Artemisinin" was presented by Tu Youyou. The foreign representatives in the audience raised many questions and participated enthusiastically in the discussions. Subsequently, the meeting split into three different groups, namely chemistry, pharmacology and toxicity, and clinical, to allow for more substantive discussions. At the end of the meeting, a development plan was approved. The WHO suggested that China set up a management and coordination agency as soon as possible to coordinate with the WHO Secretariat on implementing the plan.

The WHO Scientific Working Group on the Chemotherapy of Malaria sent their secretary, Dr Trigg, and scientific advisors, Dr Heiffer and Li Zhengjun, to China. They visited the research units and pharmaceutical factories in Beijing, Shanghai, Guilin and Guangzhou, and forged a preliminary agreement between China and the WHO regarding technical and financial support for collaborative research. Thereafter, MOH and State Drug Administration (SDA) jointly set up the National Steering Committee for Development of Artemisinin and its Derivatives (Steering Committee).

In order to provide foreign organisations with sufficient quantities of artemisinin for clinical trials and international registration, China set a number of targets for the following two years. These included attaining international standards of quality control for oral artemisinin drugs as well as artemether and artesunate injections, pre-clinical research into the pharmacology and toxicity of the drugs, and phases I, II, and III of clinical trials. At the same time, the WHO would provide staff training, equipment, and specially-bred animals for laboratory testing. Subsequently, the Scientific Working Group on the Chemotherapy of Malaria held a meeting in Geneva. The meeting gave a brief account of the research collaboration plan signed with China, and only listed the use of artesunate for the treatment of cerebral malaria as the priority project. During the meeting, concerns about the manufacturing techniques of this drug were also raised, and plans were made to send technical staff from the US Food and Drug Administration (USFDA) to China to further assess the production and management standards at the pharmaceutical factories.

At the suggestion by the WHO and with the agreement of the Chinese government, USFDA inspector Dr Tetzlaff accompanied Dr Trigg to the Kunming Pharmaceutical Factory and Guilin Pharmaceutical Factories No. 1 and No. 2 to carry out Good Manufacturing Practices

(GMP) inspections. The GMP report for the sterile workshop in Guilin Pharmaceutical Factory No. 2 where artesunate injections were manufactured was that there was a lack of stringent management in the production, the sterilisation of and testing methods did not match scientific standards, and the design of the workshop and equipment maintenance were far from ideal. They concluded that the artesunate injections manufactured by Guilin No. 2 Pharmaceutical Factory did not meet the GMP requirements and were unsuitable for clinical trials conducted outside China.

Although there was no detailed report submitted after the GMP inspection at the Kunming Pharmaceutical Factory, the USFDA personnel indicated that the problems identified were similar to those found in the Guilin Pharmaceutical Factory No. 2. In order to find a pharmaceutical factory in China that could meet the GMP requirements for the manufacture of artesunate injections, China recommended that they inspect the Shanghai Xinyi Pharmaceutical Factory which had the best production conditions in China. The report from the USFDA personnel was still the same: the production conditions did not meet GMP requirements. As a result, the cooperation between China and WHO was brought to a standstill. Dr Trigg suggested two options. One option was for China to build a new manufacturing plant which was able to fulfil the GMP requirements for the production of the artesunate injections, but this would mean a three to five-year delay in international drug registration. The second suggestion was to make use of facilities overseas to process a batch of injections that could meet the GMP requirements so that the pre-clinical pharmacology and toxicity research required for international drug registration could be completed as soon as possible.

The Steering Committee weighed the pros and cons of Dr Trigg's suggestions and decided that the best plan was to complete the international drug registration for artemisinin drug as soon as possible. Hence, it was agreed that a batch of artesunate injections would be produced overseas. At the same time, China requested that the WHO recommend a partner for the collaboration. Dr Trigg recommended the Walter Reed Army Institute of Research (WRAIR) in the US, and personally flew to the US to carry out the negotiations. The US decided to send Colonel

Brown from the International Health Affairs of the US Department of Defence and Dr Heiffer from the WRAIR to visit China to discuss the research collaboration on the artesunate injections.

The Steering Committee was concerned an agreement might not be possible due to the short notice and their inadequate understanding of the negotiation proposals from the US. Hence, they wrote to the WHO suggesting a postponement of the visit and requested that the WRAIR send the collaboration proposal to them in advance. As requested, Dr Trigg sent the Artesunate Research Collaboration Agreement, at the same time confirming that it had been approved by the WHO and US Department of Defence and hoping that China could arrange a meeting for the three parties as soon as possible. After discussion, Steering Committee felt that the collaboration was not based on the basis of "mutual benefit" and "friendly collaboration" but was an attempt to use the terms and conditions to restrict China's freedom of action. The members of the Steering Committee were not happy with the proposed terms and suggested numerous amendments.

After Dr Lucas from the WHO Special Programme for Research and Training in Tropical Diseases (TDR) came to China to discuss the Artesunate Research Collaboration Agreement, the MOH and SDA jointly submitted the agreement to senior government officials for approval. Subsequently, the agreement was approved by the authorities from both parties. Unexpectedly, several foreign research organisations began developing new drugs from artemisinin. The Swiss Roche Pharmaceuticals was able to artificially synthesise artemisinin while the WRAIR extracted artemisinin from locally-grown *A. annua* and was able to determine its physical and chemical properties. The TDR and the Dutch ACF company signed an agreement to develop artemether with a financial budget (1990–1996) of approximately US$6 million.

The Kunming and Guilin Pharmaceutical Factories continued to manufacture artesunate, artemether and other artemisinin-related products, but as the manufacturing conditions did not meet the GMP standards, the products manufactured could not be sold internationally. Even after the pharmaceutical factories and artemisinin products in China had attained the GMP standards, they could only be supplied as raw materials for the foreign pharmaceutical companies. Some of these foreign enterprises purchased artemisinin from China, then processed them into various products for sale at high prices. Others purchased semi-finished or finished artemisinin products which, after processing and re-packaging, were sold around the world for prices many times higher.

Why was there only "pain" but no "joy" in the international collaboration on artemisinin? This is a problem worthy of reflection. The source of the "pain" from the international collaboration on artemisinin was probably a result of inadequate knowledge about intellectual property rights. The collaboration involved China showing the direction and passing valuable experiences to foreign countries. As soon as they had acquired the skills and knowledge, they naturally began to commercially exploit the information. In addition, we were too concerned over the details and content of the first collaboration agreement under the assumption that we were crucial to the collaboration. As a result, we were excluded from the game even before it started. We possessed no bargaining chips other than some early experience, so we should not have been complacent but instead should have increased efforts to maintain our lead. Only then could we have been able to taken the lead in negotiations.

There is a Chinese saying "the student has outshone his master". At present, the foreign research organisations are not only capable of engaging in independent artemisinin research, but also possess an absolute advantage in other areas of artemisinin research. Even our role as supplier of artemisinin is not assured. We need to sell at a very low price or run the risk of having no market at all. The French Sanofi Company makes use of the US-patented yeast fermentation technique to produce *A. annua* acid, and the yield is astonishing. By the end of 2012, 39 tonnes had been produced, equivalent to 40 million portions of antimalarial drugs!

The 4Th Meeting of The SWG-Chemal "Qinghaosu"
Beijing · China october · 1981

The Meeting to Evaluate the Anti-Malaria Drug, Artemether, Shanghai, 1981

深入宝[illegible]采明珠

——记抗疟新药"[illegible]素"的研制历程

A special report on artemisinin, Guangming Daily, 18 June 1978

Appendix 4: Organisations Named

Name of Organisation	Chinese Name
302 Military Hospital of China	解放军第三〇二医院
Academy of Military Medical Sciences	军事医学科学院
Academy of Traditional Chinese Medicine *Current name*: *China Academy of Chinese Medical Sciences*	中医研究院中药研究所 *现名：中国中医科学院*
Beijing Institute of Materia Medica, CAMS	中国医学科学院北京药物研究所
Beijing Institute of Pharmaceutical Industry	北京制药厂工业研究所
Beijing Medical University	北京医学院
Central Military Commission	中央军事委员会
China Academy of Chinese Medical Sciences *Former name*: *Academy of Traditional Chinese Medicine*	中国中医科学院 *曾用名：中医研究院中药研究所*
China National Pharmaceutical Industry Corporation	中国医药工业公司
Chinese Academy of Medical Sciences	中国医学科学院
Chinese Academy of Sciences	中国科学院
Chinese Academy of Social Sciences	中国社会科学院
Chinese Medical Association	中华医学会
Chongqing Academy of Chinese Materia Medica	重庆中医中药研究所
Chongqing Medicine Co.	四川重庆药材公司
Chongqing Pharmaceutical Industry	重庆医药工业

Appendix 4: (*Continued*)

Name of Organisation	Chinese Name
Chongqing Xinan Pharmaceutical Factory	重庆西南制药厂
Commission for Science, Technology and Industry for National Defense	中国人民解放军国防科学技术委员会
Ding Xiang Yuan Chinese Medicine Forum	丁香园中医版
Epidemic Prevention Squad, Hainan Military District	海南军区防疫队
Fourth Military Medical University	第四军医大学
Fujian Province Chinese Medical Research Institute *Current name*: *Fujian University of Traditional Chinese Medicine*	福建省中医药研究院 *现名：福建中医药大学*
General Logistics Department, People's Liberation Army	中国人民解放军总后勤部
Guang'anmen Hospital, CACMS	中国中医科学院北京广安门医院
Guangdong Academy of Chinese Medicine	广东中医学院
Guangdong Institute of Chinese Medicine	广东中药研究所
Guangdong Institute of Parasitic Diseases	广东寄生虫病研究所
Guangzhou University of Chinese Medicine	广州中医学院
Guangzhou Zhongshan School of Medicine	广州中山医学院
Guilin Pharmaceutical Factory No. 1	广西桂林第一制药厂
Guilin Pharmaceutical Factory No. 2	广西桂林第二制药厂
Hainan General Hospital	海南人民医院

Hainan Institute of Parasitic Diseases	海南寄生虫病研究所
Hainan Military District Hospital No. 187	海南军区187医院
Hangzhou Pingfeng Mountain Workers' Sanatorium	杭州屏峰山工人疗养院
Health and Epidemic Prevention Institute, Logistics Department, Guangzhou Military District	广州军区后勤部卫生防疫研究所
Health Bureau, Gaoyou County, Jiangsu Province	江苏省高邮县卫生局
Health Institute, Logistics Department, Nanjing Military District	南京军区后勤部卫生部
Henan Pharmaceutical Factory	河南制药厂
Henan Provincial Institute of Food and Drug Control	河南省药品检定所
Huadong Institute of Entomology, CAS	中国科学院华东昆虫研究所
Hubei Jianmin Pharmaceutical Factory	湖北健民制药厂
Institute of Biophysics, CAS	中国科学院生物物理研究所
Institute of Botany of Jiangsu Province, CAS	中国科学院江苏省植物研究所
Institute of Chinese Materia Medica, CACMS	中国中医科学院中药研究所
Institute of Health and Epidemic Prevention, Guangzhou Military District	广州军区卫生防疫研究所
Institute of Materia Medica, CAMS	中国医学科学院药物研究所
Institute of Materia Medica, CAS	中国科学院药物研究所
Institute of Military Medical Sciences, Guangzhou Military District	广州军区军事医学研究所
Institute of Military Medical Sciences, Kunming Military District	昆明军区军事医学研究所

Appendix 4: (*Continued*)

Name of Organisation	Chinese Name
Institute of Military Medical Sciences, Nanjing Military District	南京军区军事医学研究所
Institute of Parasitic Diseases, Hubei Academy of Medical Sciences	湖北省医学科学院寄生虫病研究所
Institute of Plant Physiology and Ecology, Shanghai Institutes for Biological Sciences, CAS	中国科学院上海生命科学研究院植物生理生态研究
Institute of Zoology, CAS	中国科学院动物研究所
Jiangsu Institute of Parasitic Diseases	江苏省血防所
Jiangsu Institute of Traditional Chinese Medicine	江苏中医研究所
Kunming Institute of Botany, CAS	中国科学院昆明植物所
Kunming Institute of Chinese Medicine	昆明中药研究所
Kunming Pharmaceutical Factory	云南昆明制药厂
Logistics Health Office, Hainan Military District	海南军区后勤卫生处
Malaria Committee, MOH	卫生部疟疾专题委员会
Military Control Commission	军事管制委员会（军管会）:
Military Control Commission, MOH	卫生部军事管制委员会
Military Medical Institute, Logistics Department, Kunming Military District	昆明军区后勤部军事医学科学研究所
Ministry of Chemical Industry	化工部
Ministry of Health	卫生部

Nanjing Institute of Botany, CAS	中国科学院南京植物研究所
Nanjing Military District Hospital No. 81	南京军区八一医院
Nanjing Pharmaceutical Factory	南京制药厂
Nanjing University of Chinese Medicine	南京中医药大学（南京中医学院）
National Institute of Parasitic Diseases, CAMS	中国医学科学院寄生虫研究所
National Institutes for Food and Drug Control	中国药品生物制品检验所
National Steering Committee for Development of Artemisinin and Its Derivatives	中国青蒿素及其衍生物研究开发指导委员会
National Vaccine and Serum Institute	北京生物制品研究所
Northeast Pharmaceutical Factory	东北制药厂
Northern Jiangsu People's Hospital	苏北人民医院
Peking Anti-Imperialist Hospital *Current name: Peking Union Medical College Hospital*	北京反帝医院 *现名：北京协和医院*
Peking Union Medical College Hospital *Former name: Peking Anti-Imperialist Hospital*	北京协和医院 *曾用名：北京反帝医院*
People's Liberation Army	中国人民解放军
Revolutionary Committee Professional Group, CACMS	中国中医科学院革委会业务组
Science and Technology Division, Ministry of Health	卫生部科技司
Science and Technology Planning Department, AMMS	军事医学科学院科技部计划处
Second Military Medical University	第二军医大学

Appendix 4: (*Continued*)

Name of Organisation	Chinese Name
Shandong Academy of Chinese Medicine	山东省中医药研究所
Shandong *Artemisia Annua* L. Research Collaboration Team	山东省黄花蒿研究协作组
Shandong Institute for Drug Control *Current name: Shandong Institute for Food and Drug Control*	山东药品检验所 *现名：山东省食品药品检验研究院*
Shandong Institute of Parasitic Diseases	山东寄生虫病研究所
Shandong Institute of Scientific and Technical Information	山东省科技情报所
Shandong Provincial Academy of Integrative Medicine	山东省中西医结合研究院
Shanghai Institute for Food and Drug Control	上海市卫生局药品检验所 （上海市食品药品检验所）
Shanghai Institute of Materia Medica, CAMS	中国医学科学院上海药物研究所
Shanghai Institute of Materia Medica, CAS	中国科学院上海药物研究所
Shanghai Institute of Occupational Health	上海劳动卫生职业病研究所
Shanghai Institute of Organic Chemistry	中国科学院上海有机所
Shanghai Institute of Parasitic Diseases	上海寄生虫病研究所
Shanghai Institute of Pharmaceutical Industry	上海医药工业研究院
Shanghai Institutes for Biological Sciences, CAS	中国科学院上海生命科学研究院
Shanghai Literature Institute of Traditional Chinese Medicine	上海中医文献研究馆

Shanghai Pharmaceutical Factory No. 10	上海第十制药厂
Shanghai Pharmaceutical Factory No. 11	上海第十一制药厂
Shanghai Pharmaceutical Factory No. 14 *Current name*: *Shanghai Zhongxi Pharmaceutical Co. Ltd.*	上海第十四制药厂 *现名：上海中西药业股份有限公司*
Shanghai Pharmaceutical Factory No. 2	上海第二制药厂
Shanghai Pharmaceutical Industry	上海医药工业
Shanghai Pharmaceutical Industry Research Institute	上海医药工业研究院
Shanghai Pharmaceuticals Holding Co. Ltd.	上海医药集团股份有限公司
Shanghai Plastics Factory No. 2	上海塑料二厂
Shanghai Research Institute of Acupuncture	上海针灸研究所
Shanghai Xinyi Pharmaceutical Factory	上海信谊制药厂
Shanghai Zhongzhou Pharmaceutical Factory	上海中州药厂
Shenyang Pharmaceutical Industry Branch Office	沈阳医药工业分公司
Sichuan Institute of Chinese Medicine	四川省中药研究所
Sichuan Wuling Shan Pharmaceutical Factory	四川省武陵山制药厂
Sichuan Academy of Chinese Medicine Sciences	四川中药研究所
South China Institute of Botany, CAS *Current name: South China Botanical Garden, CAS*	中国科学院华南植物研究所 *现名：中国科学院华南植物园*
State Council Science and Education Group	国务院科教组

Appendix 4: (*Continued*)

Name of Organisation	Chinese Name
State Drug Administration	国家医药管理总局
Current name: China Food and Drug Administration	*现名：国家食品药品监督管理总局*
State Science and Technology Commission	科学技术委员会
Current name: Ministry of Science and Technology	*现名：科学技术部*
Third Military Medical University	第三军医大学
Tianjin Pesticide Experimental Plant	天津农药实验场
Tianjin Wuqing Chinese Medical Hospital	天津武清中医院
Xi'an Pharmaceutical Factory	西安制药厂
Yunnan Institute of Materia Medica	云南省药物研究所
Yunnan Institute of Parasitic Diseases	云南省寄生虫病防治所
Zhuzhuang Brigade, Guandong Commune, Juye County, Shandong	山东巨野县城关东公社朱庄大队

Appendix 5: List of Plants

Plants	Chinese Name
Aconitum	乌头
Agrimonia pilosa Ledeb.	仙鹤草
Ailanthus altissima	臭椿
Aristolochia yunnanensis Franch.	云南马兜铃
Artabotrys hexapetalus	鹰爪
Artemisia absinthium	苦蒿
Artemisia annua	青蒿
Artemisia annua L.	黄花蒿
Artemisia annua L. f. *macrocephala* Pamp.	大头黄花蒿
Brucea javanica	鸦胆子
Cortex lycii	地骨皮
Eleocharis dulcis	菱花
Euphorbia helioscopia	五朵云
Euphorbia kansui	甘遂
Fructus mume	乌梅
Hemerocallis citrina Baroni	黄花
Hydrangea macrophylla	绣球
Hypericum japonicum Thunb.	地耳草
Nandina domestica	南天竹
Piper nigrum	胡椒
Polyalthia nemoralis	陵水暗罗
Radix kansui	甘遂
Strigose Hydrangea	伞花八仙
Verbena officinalis L.	马鞭草

Tu Youyou's Nobel Lecture[1]

7 Dec 2015

Artemisinin– A Gift from Traditional Chinese Medicine to the World

Tu Youyou

Institute of Chinese Materia Medica, China Academy of Chinese Medical Sciences, Beijing, 100700, China

Nobel Lecture, Dec 7, 2015
Karolinska Institutet

Slide 1

Source: Nobelprize.org (*http://www.nobelprize.org/nobel_prizes/medicine/laureates/2015/tu-lecture.html*).

Dear respected Chairman, General Secretary, esteemed Nobel laureates, ladies and gentlemen,

It is my great honour to give this lecture today at Karolinska Institutet. The title of my presentation is: Artemisinin — A Gift from Traditional Chinese Medicine to the World.

Before I start, I would like to thank the Nobel Assembly and the Nobel Foundation for awarding me the 2015 Nobel Prize in Physiology or Medicine. This is not only an honour for myself, but also a recognition and motivation for all scientists in China. I would also like to express my sincere appreciation for the gracious hospitality of the Swedish people which I have received during my short stay over the last few days. Thanks also to Dr William Campbell and Dr Satoshi Omura for their excellent and inspiring presentations.

The story I will tell today is about the diligence and dedication of Chinese scientists during the search for anti-malarial drugs from traditional Chinese medicines 40 years ago under considerably under-resourced research conditions.

Discovery of Artemisinin

Some of you may have read the history of the discovery of artemisinin in numerous publications. I will give a brief review here. This slide (Slide 2) summarises the anti-malaria research programme carried out by the team at the Institute of Chinese Materia Medica (ICMM) of the Academy of Traditional Chinese Medicine (ATCM). The programmes highlighted in blue were accomplished by the team at ATCM while the programmes highlighted in blue and white were completed through joint efforts by the teams at ATCM and other institutes. The other programmes were completed collaboratively by other research teams across the nation.

Discovery of Artemisinin at ICMM

The team at ICMM initiated the research on Chinese medicines for malaria treatment in 1969. Following extensive screening, we started to

Summary of the Work Completed by the Research Team in Academy of Traditional Chinese Medicine (Boxes in Blue Background)

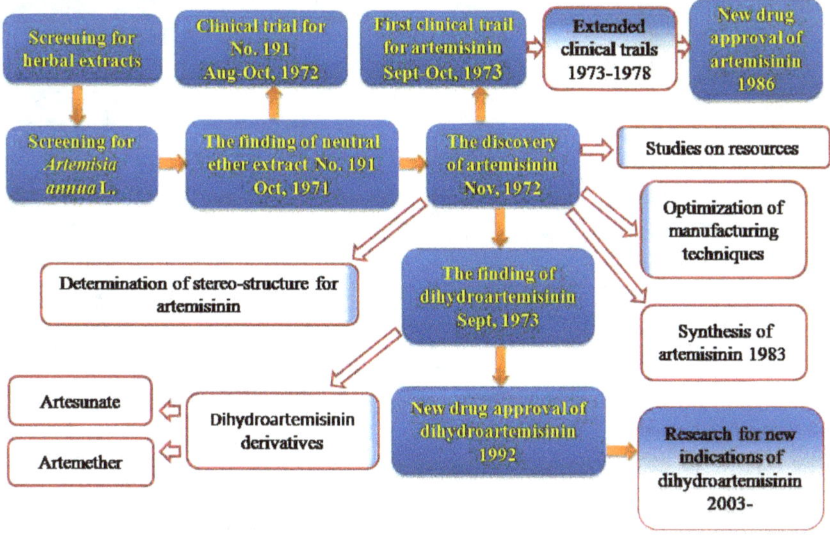

Slide 2

focus on the herb *qinghao* (*Artemisia annua*) in 1971, but received no promising results after multiple attempts. In September 1971, a modified procedure was designed to reduce the extraction temperature by immersing or distilling *qinghao* using ethyl ether. The extracts obtained were then treated with an alkaline solution to remove acidic impurities and retain the neutral portion. In the experiments carried out on 4 October 1971, sample No. 191, i.e. the neutral portion of the *qinghao* ether extract, was found to be 100% effective on malaria-infected mice when administered orally at a dose of 1.0g/kg for three days consecutively (Slide 3). The same results were observed when tested on malaria-infected monkeys between December 1971 and January 1972. This breakthrough finding became a critical step in the discovery of artemisinin.

We subsequently carried out a clinical trial between August and October 1972 in Hainan province in which the neutral *qinghao* ether extract successfully cured thirty *falciparum* and *Plasmodium* malaria patients. This was the first time the neutral *qinghao* ether extract was

The Breakthrough in the Research of *Artemisia annua* L for Discovery of Artemisinin

Archive, Institute of Chinese Materia Medica

Slide 3

tested on humans. In November 1972, an effective anti-malarial compound was isolated from the neutral *qinghao* ether extract. The compound was later named *qinghaosu* (artemisinin in Chinese).

Artemisinin Chemistry Studies

We started to determine the chemical structure of artemisinin in December 1972 through elemental analysis, spectrophotometry, mass spectrometry, polarimetric analysis and other techniques. These experiments confirmed that the compound had a completely new sesquiterpene structure with a formula of $C_{15}H_{22}O_5$, a molecular weight of 282, and contained no nitrogen (Slide 4).

Stereo-Structure of Artemisinin

The formula of the molecule and other results were verified by the Analytical Chemistry Department of the China Academy of Medical

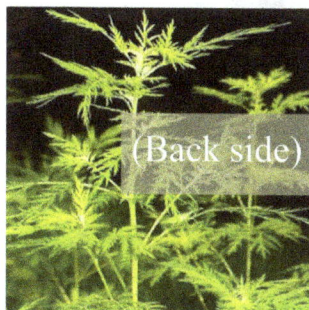

中国医学科学院药物研究所
分析化学室
微量有机元素分析报告

样品原编号：青蒿素(II)
分析编号：
送样人：中药研究所屠呦呦
工作領导人签字：
样品乾燥情况：已干燥
样品主要性质：
感光＿＿，吸水＿＿，挥发性＿＿，爆炸性＿＿
沸点＿＿，熔点＿＿，其 他＿＿

样品含有元素：
要求分析项目：C，H，O
要求分析元素的百分比范围或化合物的可能結构式：$C_{15}H_{22}O_5$
送样日期：73年4月24日

(分析結果見

The elements analysis by the collaborative institution, Institute of Matria Medica, Chinese Academy of Medical Sciences, on April 27th, 1973. (Front side)

(Back side)

分析結果：　　样品在分析前的处理：

样品重(Mg)	CO_2重(Mg)	H_2O重(Mg)	%C	%H	註
2.531	5.923	1.928	63.86	8.52	
3.173	7.411	2.422	63.59	8.54	
样品重(Mg)	温度(°C)	压力(MM/Hg)	氮气体积(Ml)	%N	
样品重(Mg)	(　)沉淀重(Mg)	Ml	N	%	

分析室意見：
分析人：[illegible]　　分析日期 73 年 4 月 27 日

Archive, Institute of Chinese Materia Medica

Slide 4

Sciences on 27 April 1973. We began a collaboration with the Shanghai Institute of Organic Chemistry and the Institute of Biophysics of the Chinese Academy of Sciences on the analysis of the chemical structure of artemisinin in 1974. The stereo-structure was finally determined using the X-ray diffraction technique which verified that artemisinin was a new sesquiterpene lactone containing a peroxyl group (Slide 5). The structure was published in 1977[2] and cited in *Chemical Abstracts.*[3]

Artemisinin and Artemisinin Derivatives

Derivatisation of artemisinin was performed in 1973 in order to determine its functional group. A carboxyl group was verified in the

---

[2] Research Group for Artemisinin. "A new sesquiterpene lactone — Artemisinin". *Chinese Science Bulletin*, 1977, 3:142.
[3] C.A. 1977, 87, 98788g.

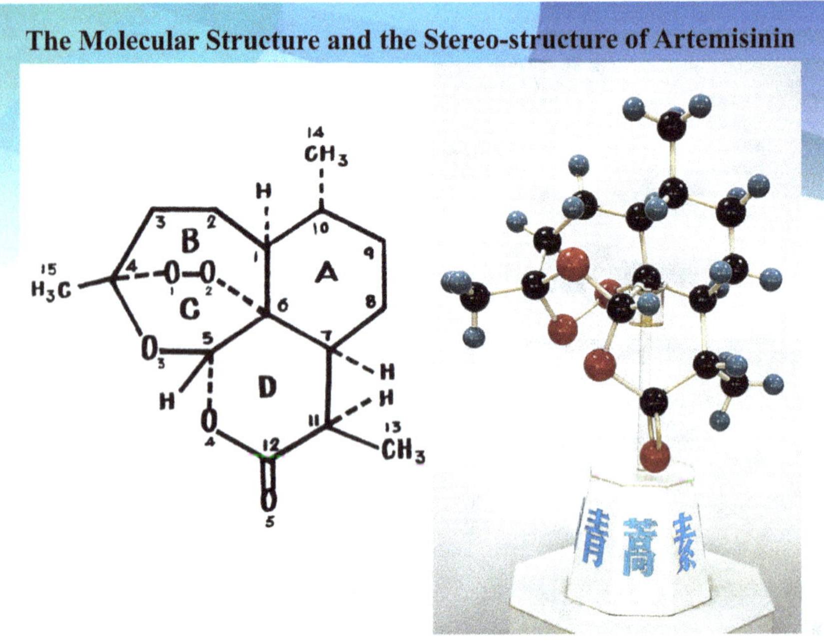

The Molecular Structure and the Stereo-structure of Artemisinin
青蒿素

Slide 5

artemisinin molecule through reduction by sodium borohydride. Dihydroartemisinin was found in this process. Further research on the structure-activity relationship of artemisinin was conducted (Slides 6, 7). The peroxyl group in the artemisinin molecule was proven critical for its anti-malarial function. Efficacy was improved for some compounds derivatised through the hydroxyl group of dihydroartemisinin.

Artemisinin and Artemisinin Derivatives

This slide (Slide 8) shows the chemical structures of artemisinin and its derivatives — dihydroartemisinin, artemether, artesunate and arteether. Up to now, no clinical application has been reported with other artemisinin derivatives except for the four presented here.

Structure - Activity Relationship of Compounds from Artemisinin

Compound	Graph of chemical structures	Dose mg/kg/day × 3	Clearance of parasites
Artemisinin	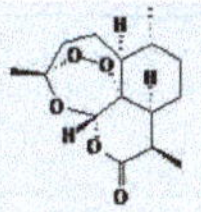	50-100	Yes
Dihydroartemisinin	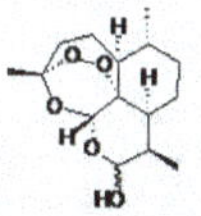	12.5	Yes
Acetate of dihydroartemisinin	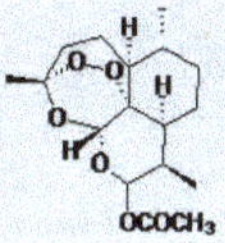	6	Yes
Deoxyartemisinin	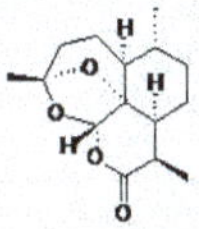	100	No

Slide 6

Structure - Activity Relationship of Compounds from Artemisinin (continued)

Compound	Graph of chemical structures	Dose mg/kg/day × 3	Clearance of parasites
Dihydrodeoxyartemisinin	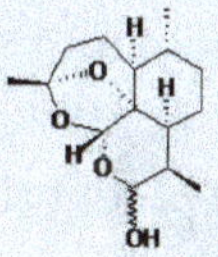	100	No
Acid treatment product of artemisinin	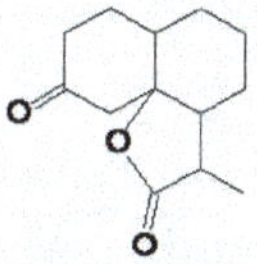	100	No
Base treatment product of artemisinin	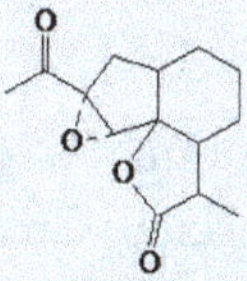	100	No

Slide 7

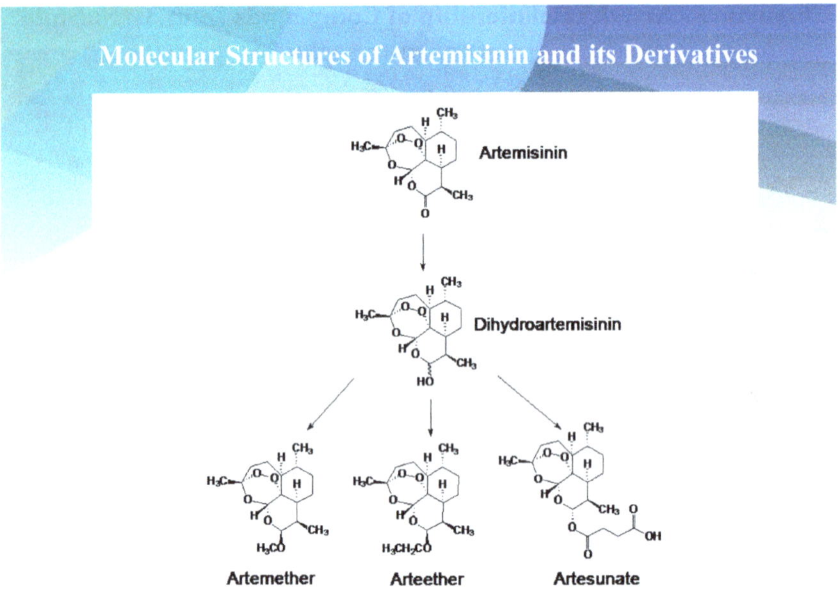

Slide 8

Artemisinin and Dihydroartemisinin New Drug Certificates

This slide (Slide 9) shows the Artemisinin New Drug Certificate (left) and the Dihydroartemisinin New Drug Certificate (right) granted by the China Ministry of Health in 1986 and 1992 respectively. Dihydroartemisinin is ten times more potent than artemisinin, and again demonstrated the high efficacy, rapid action and low toxicity of the drugs in the artemisinin category.

Worldwide Attention on Artemisinin

The World Health Organization (WHO), the World Bank and the United Nations Development Program (UNDP) held the 4th Joint Malaria Chemotherapy Science Working Group meeting in Beijing in

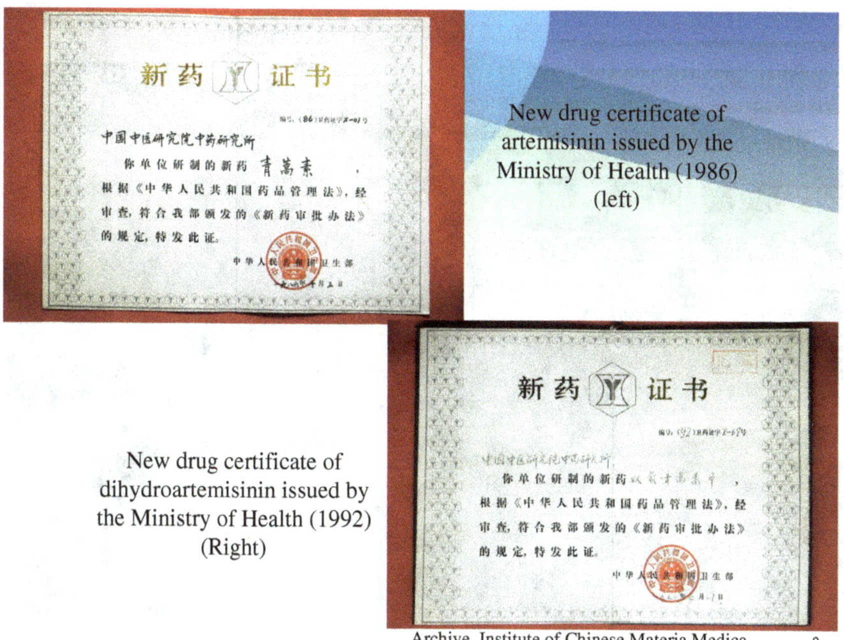
新药 证书
中国中医研究院中药研究所
你单位研制的新药 青蒿素
根据《中华人民共和国药品管理法》，经审查，符合我部颁发的《新药审批办法》的规定，特发此证。
中华人民共和国卫生部
New drug certificate of artemisinin issued by the Ministry of Health (1986) (left)
新药 证书
根据《中华人民共和国药品管理法》，经审查，符合我部颁发的《新药审批办法》的规定，特发此证。
中华人民共和国卫生部
New drug certificate of dihydroartemisinin issued by the Ministry of Health (1992) (Right)
Archive, Institute of Chinese Materia Medica
9

Slide 9

1981 (Slide 10). A series of presentations on artemisinin and its clinical application including my report *Studies on the Chemistry of Qinghaosu* received positive and enthusiastic responses. In the 1980s, several thousand malaria patients were successfully treated with artemisinin and its derivatives in China.

After this brief review, you may comment that this is no more than an ordinary drug discovery process. However, it was not a simple and easy journey in the discovery of artemisinin from *qinghao*, a Chinese herbal medicine with over two thousand years of clinical application.

Commitment to the Clearly-Defined Goal Assures Success in Discovery

The Institute of Chinese Materia Medica of the Academy of Traditional Chinese Medicine joined the national "523" anti-malaria

Worldwide Attention to Artemisinin
The 4Th Meeting of The SWG-Chemal "Qinghaosu"
Beijing · China october·1981

Slide 10

research project in 1969. I was appointed the head by the academy's leadership team and tasked to build the "523" research group in the Institute, responsible for developing new anti-malaria drugs from Chinese medicines. It was a high priority, confidential military programme. As a young scientist early in her career, I felt overwhelmed by the trust and responsibility for such a challenging and critically important task. I felt impelled to fully devote myself to accomplishing my duties (Slide 11).

Knowledge is a Prologue in Discovery

This is a photo taken soon after I joined the Institute of Chinese Materia Medica (Slide 12). Professor Lou Zhiqin, a famous pharmacologist, was mentoring me on how to differentiate herbs. I attended a training course on the theories and practices of traditional Chinese medicine

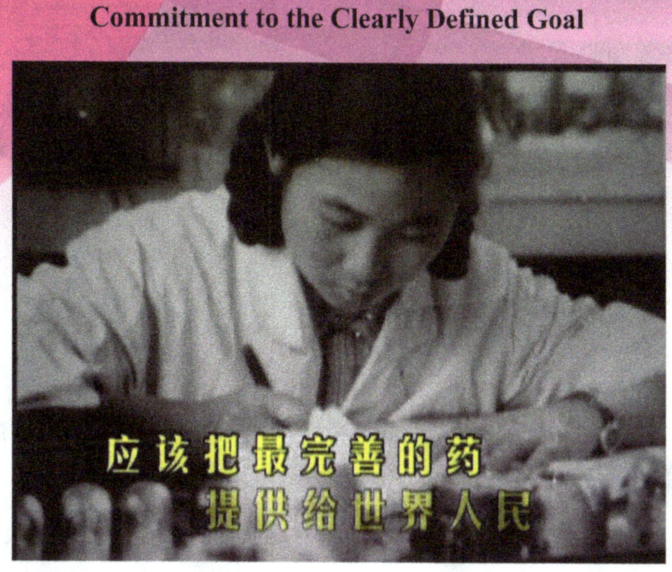

Commitment to the Clearly Defined Goal
应该把最完善的药
提供给世界人民

Slide 11

designed for professionals with a background in modern (western) medical training. "Fortune favours the prepared mind" and "what is past is prologue". My prologue of integrated training in modern and Chinese medicine prepared me for the challenges when the opportunities in searching for anti-malarial Chinese medicine became available.

Information Collating and Accurate Deciphering Are the Foundation for Success in Research

After receiving the tasks, I collected over 2,000 herbal, animal and mineral prescriptions for either internal or external use through reviewing ancient traditional Chinese medical literature and folk recipes and interviewing well-known and experienced Chinese medical physicians who provided me with prescriptions and herbal recipes. I summarised 640 prescriptions in the brochure *Collection of Anti-Malarial Recipes*

Slide 12

and Prescriptions (Slide 13). It was this information collection and deciphering that laid a sound foundation for the discovery of artemisinin. This also differentiates the approaches taken by Chinese medicine and general phytochemistry in the search for novel drugs.

Literature Review Inspires an Idea Leading to Success

I reviewed the traditional Chinese literature again when our research stalled following numerous failures. In reading Ge Hong's *A Handbook of Prescriptions for Emergencies* (Eastern Jin Dynasty, 3rd to 4th century) (Slide 14) I further considered the sentence "A handful of *qinghao* immersed in two litres of water, wring out the juice and drink it all" when *qinghao* was mentioned for alleviating malaria symptoms. This suggested to me that heating might need to be avoided during

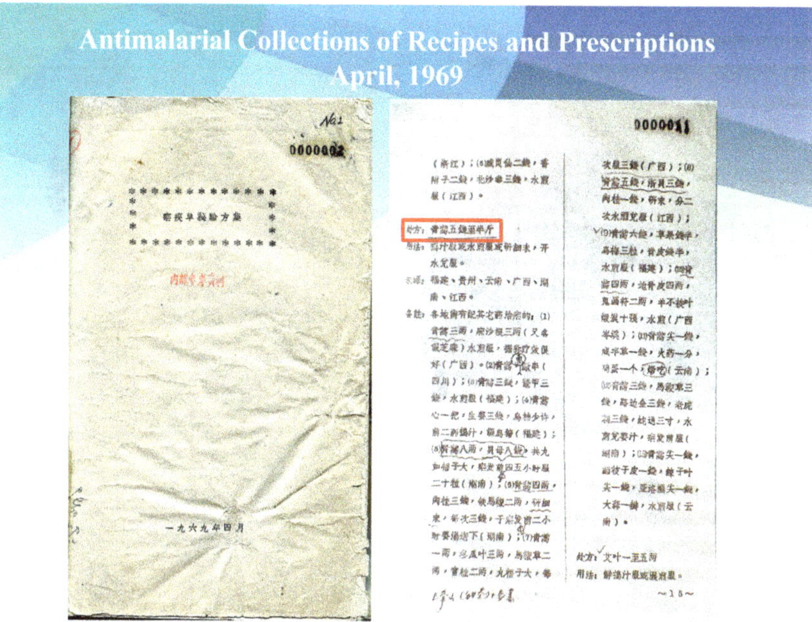

Archive, Institute of Chinese Materia Medica

Slide 13

extraction, therefore the method was modified by using a solvent with a low boiling point.

The earliest mention of *qinghao's* application as a herbal medicine was found on the silk manuscripts entitled *Prescriptions for Fifty-two Kinds of Disease* unearthed from the third Han Tomb at Mawangdui. Its medical application was also recorded in *Shen Nong's Herbal Classic*, *Buyi Leigong Baozhi*, *Compendium of Materia Medica* etc. However, no clear classification was given for *qinghao* (*Artemisia*) despite several mentions of the name "*qinghao*" in the literature (Slide 15). All species of the *qinghao* (*Artemisia*) family were mixed together and by the 1970s, two *qinghao* (*Artemisia*) species were recorded in *Chinese Pharmacopoeia* and four others were also being prescribed. Our subsequent investigation proved that only *Artemisia annua* L. contains artemisinin and is effective against malaria.

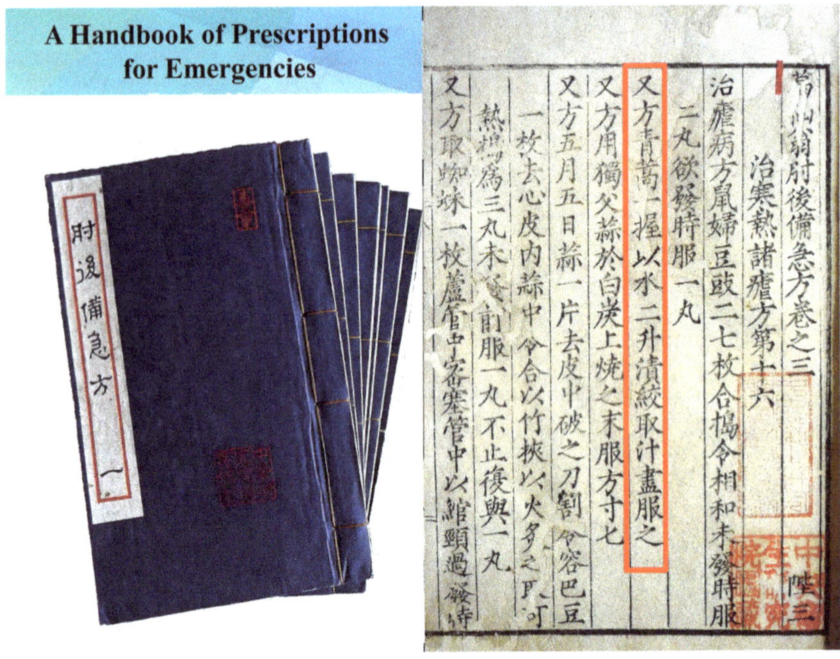

Slide 14

In addition to the confusion in the selection of the right species and the difficulty caused by the low content of artemisinin in the herb, variables such as identifying the medicinal part of the plant, the growing regions, the harvest season, and extraction/purification processes etc. added extra difficulties to the discovery of artemisinin. Success in identifying the effectiveness of the *qinghao* neutral ether extract was not achieved easily.

No doubt, traditional Chinese medicine provides a rich resource. Nevertheless, it requires thoughtful consideration to explore and improve it.

Persistence in the Face of Challenges

Research conditions were relatively poor in China in the 1970s (Slide 16). In order to produce sufficient quantity of *qinghao* extract for clinical trials, the team carried out extraction using several household water vats. Some members' health deteriorated due to exposure to the large

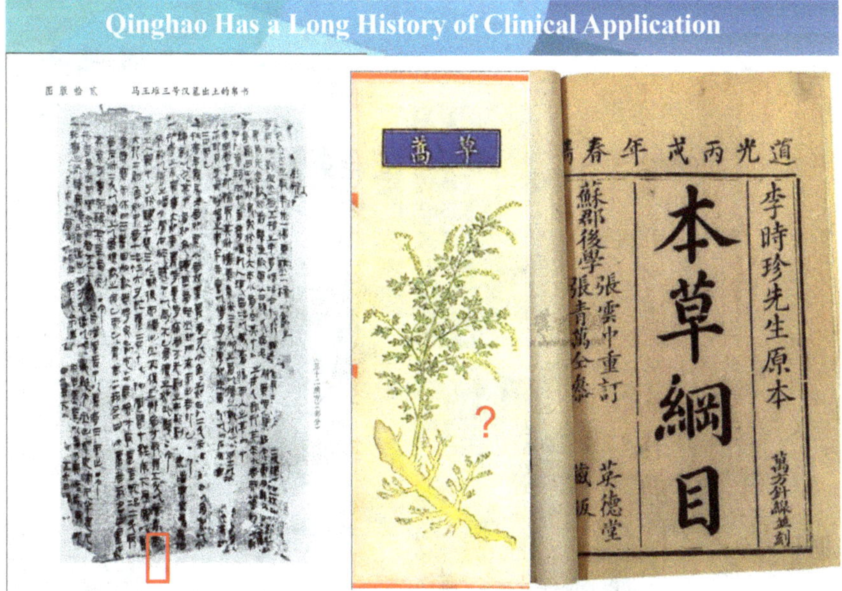

Slide 15

quantity of organic solvent and insufficient ventilation equipment. In order to launch clinical trials sooner while not compromising patient safety, based on the limited safety data from the animal studies, the team members and myself volunteered to take *qinghao* extract ourselves to assure its safety.

Unsatisfactory results were observed in the clinical trial using artemisinin tablets. The team carried out a thorough investigation and verified that poor disintegration of the tablets was the root cause. This allowed us to quickly resume the trial using capsules and confirm artemisinin's clinical efficacy in time.

Collaborative Team Efforts Expedited Translation from Scientific Discovery to Effective Medicine

An anti-malaria drug research symposium was held by the National Project 523 Office in Nanjing on 8 March 1972. At this meeting, on

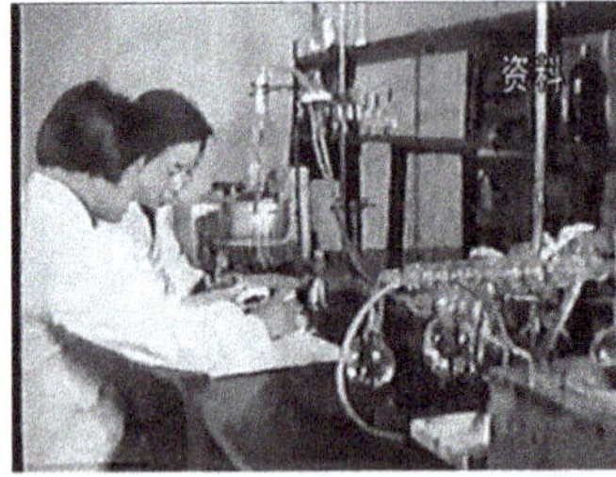

Slide 16

behalf of the Institute of Chinese Materia Medica, I reported the positive results of the *qinghao* extract No. 191 observed in animal studies performed on malaria-infected mice and monkeys. The presentation received significant interest. On 17 November 1972, I reported the successful treatment of 30 clinical cases at a national conference held in Beijing. This triggered nationwide collaboration in research on *qinghao* for malaria treatment.

Today, I would like to express my sincere appreciation again to my fellow Project 523 colleagues in the Academy of Traditional Chinese Medicine for their devotion and exceptional contributions during the discovery and subsequent application of artemisinin. I would like to, once again, thank and congratulate the colleagues from Shandong Provincial Institute of Chinese Medicine, Yunnan Provincial Institute of Materia Medica, the Institute of Biophysics of the Chinese Academy of Sciences, Guangzhou University of Chinese Medicine, the Academy of Military Medical Sciences and many other institutes for their invaluable contributions in their respective areas of responsibility during the collaboration and for their help in caring for the malaria patients. I would also like to express my sincere respect to the National 523 Office leadership team for their continuous efforts in organising and coordinating the anti-malaria research programmes.

Without these collective efforts, we would not be able to present artemisinin — our gift to the world — in such a short period of time (Slide 17).

Malaria Remains a Severe Challenge to Global Public Health

"The findings in this year's World Malaria Report demonstrate that the world is continuing to make impressive progress in reducing malaria cases and deaths," Dr Margaret Chan, Director-General of the WHO, commented in the recent World Malaria Report (Slide 18). Nevertheless,

Invention Certificate for Progress in Anti-malarial Research
Issued by National Congress of Science and Technology, 1978
Archive, Institute of Chinese Materia Medica

Slide 17

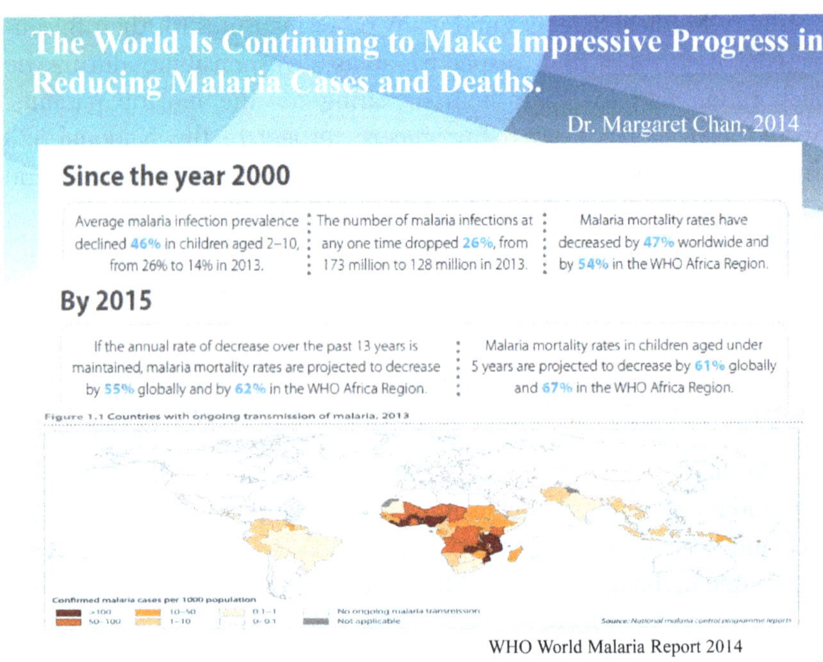

Slide 18

statistically, there are approximately 3.3 billion people in 97 countries or regions still at a risk of contracting malaria, and around 1.2 billion people live in high risk regions where the infection rate is as high as or over 1/1,000.[4]

According to the latest statistical estimate, in 2013, approximately 198 million cases of malaria occurred globally causing 580,000 deaths. 90% of them were from severely affected African countries of which 78% were children below the age of five. Only 70% of malaria patients receive artemisinin combination therapies (ACTs) in Africa and as many as 56 million to 69 million child malaria patients do not have ACTs available to them.[5]

---

[4]WHO World Malaria Report 2014.
[5]WHO World Malaria Report 2014.

The Severe Warning of Parasites Resistant to Artemisinin

P. falciparum resistance to artemisinin has been detected in five countries of the Greater Mekong sub-region: Cambodia, the Lao People's Democratic Republic, Myanmar, Thailand and Vietnam. In many areas along the Cambodia–Thailand border, *P. falciparum* has become resistant to most available anti-malarial medicines. This slide (Slide 19) shows the areas where artemisinin-resistant malaria appears according to this year's report. Artemisinin-resistant *P. falciparum* has been detected in the areas highlighted in red and black. Clearly, this is a severe warning since the resistance to artemisinin has been detected not only in the Greater Mekong sub-region but also in some African regions.

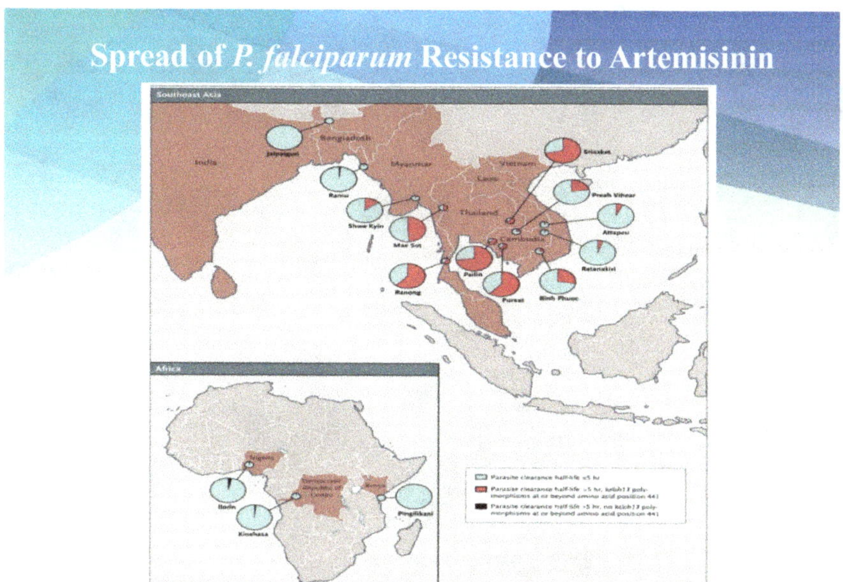

Ashley E A et al, Spread of Artemisinin Resistance in *Plasmodium falciparum* Malaria *N Engl J Med.* 2014, 371(5): 411–423

Slide 19

Global Plan for Artemisinin Resistance Containment

The goal of the Global Plan for Artemisinin Resistance Containment[6] (GPARC) is to protect ACTs as an effective treatment for *P. falciparum* malaria (Slide 20). Artemisinin resistance has been confirmed within the Greater Mekong sub-region, and potential epidemic risk is under a critical review. A unanimous agreement has been reached by over a hundred experts involved in the programme that the probability of containing and eradicating artemisinin-resistant malaria is very limited and there is an urgent need to constrain artemisinin resistance.

To protect the efficacy of ACTs, I strongly urge a global compliance to the GPARC. This is our responsibility as a scientists and medical doctors in the field.

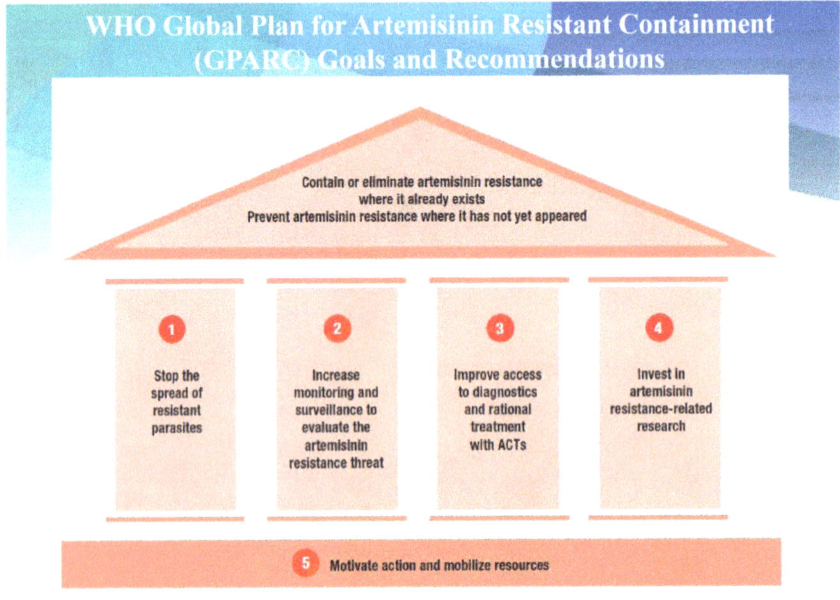

Slide 20

[6] WHO Global Plan for Artemisinin Resistance Containment, 2011.

Chinese Medicine, A Great Treasure

Before I conclude today's lecture, I would like to briefly discuss Chinese medicine. "Chinese medicine and pharmacology are a great treasure-store. We should explore them and raise them to a higher level." (Slide 21) Artemisinin was explored from this resource. From our research experience in the discovery of artemisinin, we learnt strengths from both Chinese and Western medicines. There is great potential for future advances if these strengths can be fully integrated. We have substantial natural resources from which our fellow medical researchers can develop novel medicines. Since "tasting of a hundred herbs by Shen Nong", Chinese medicine has accumulated substantial experience in clinical practice, and integrated and summarised the medical applications of many natural resources over thousands of years. Adopting, exploring, developing and advancing these practices would allow us to discover more novel medicines beneficial to world healthcare.

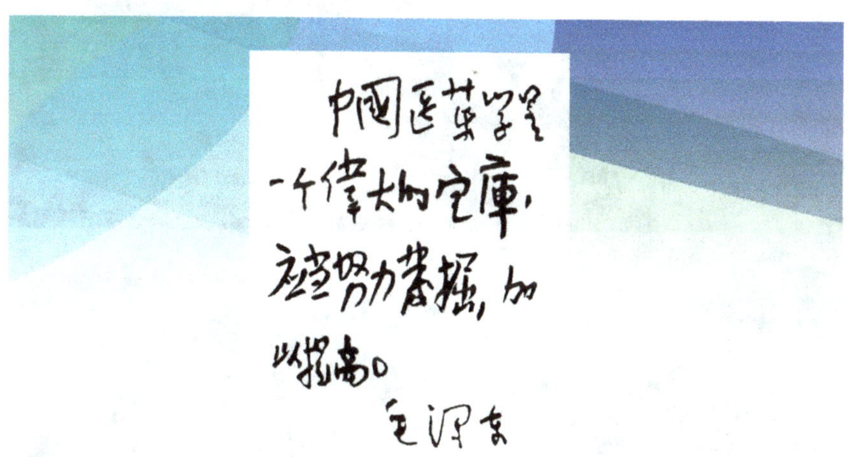

Chinese medicine and pharmacology are a great treasure-house.
We should explore them and raise them to a higher level

Hand writing by Mao Zedong

Slide 21

On the Stork Tower

Finally, I would like to share with you the well-known Tang Dynasty poem by Wang Zhihuan, *On the Stork Tower* (Slide 22).

On the Stork Tower[7]
The sun along the mountain bows;
The Yellow River seawards flows;
You will enjoy a grander sight,
By climbing to a greater height.

Let us reach to greater heights to appreciate Chinese culture and find beauty and treasure in traditional Chinese medicine!

Slide 22

[7] Wang, Zhihuan (688–742 AD). The Stork Tower is located in Yongji County, Shanxi Province.

Acknowledgements

Finally, I would like to acknowledge all colleagues in both China and overseas for their contributions in the discovery, research and clinical application of artemisinin.

I am grateful for all my family members for their continuous understanding and support.

I sincerely appreciate your kind attention.

Thanks you all!

Slide 23

Editor's Epilogue

In the 120-year history of the Nobel Prize, this was the first time that it was awarded to a female Chinese scientist, the first time a Prize in the natural sciences was awarded to a home-grown Chinese scientist, and the first time a home-grown Chinese scientist was awarded the Prize in Physiology or Medicine.

On 5 October 2015, at 5.30pm (Beijing time), Tu Youyou created three firsts in the history of the Nobel Prize!

At this time, the first thought that came to our mind was the need to do something to mark this event.

Professors Rao Yi and Zhang Daqing have taught about the history of artemisinin research, and their articles published in various media have had great impact. After receiving the news, we immediately established contact with the three authors. We are very grateful for their generous support and confidence, and we have seen the altruism underlying their perspectives on scientific research. In particular, Ms Li Runhong sacrificed precious time with her son in order to do this work.

Due to time constraints, we were not able to edit the three teachers' academic writing into a version more suited for the general public, and some of the notes and references in the original text have been omitted. In time, we will post additional content on our company's WeChat platform and our official website for readers who wish to read more.

With regard to this Nobel Prize, other than acclaim and analysis which appeared in the media, we wish to present a deeper understanding of the events of the great scientific collaboration that took place half a century ago. Professor Rao Yi and the other two authors provide an objective and independent evaluation of the scientific research leading up to the Nobel Prize.

Professor Zeng Qingping of the Guangzhou University of Chinese Medicine has been involved in artemisinin research for many years and has written numerous articles, many available on the *ScienceNet.cn* website. Professor Zeng not only gave us permission to use his articles but also wrote vividly of his continuing research on the use of artemisinin. This he completed after working rapidly for several days.

Mr Zhang Tiankan, a notable writer on scientific topics who has followed closely the news of the Nobel Prize in Physiology or Medicine, gave us permission to use and edit his articles. We sincerely thank him for his trust and support.

In the years of wondering when a home-grown Chinese scientific researcher would receive the Nobel Prize, no one guessed that this scientific research performed several decades ago would be the first. There is a reason for every event. In the same way, this book has brought together many friends, colleagues and members of the media who have helped us along the way. We are very grateful to them all.

Due to time constraints, omissions are inevitable. We look forward to receiving comments and feedback from our readers.

Our public WeChat username: *cspbooks*
Our official website: *www.cspbooks.com.cn*

The Editor
October 2015

www.ingramcontent.com/pod-product-compliance
Lightning Source LLC
Chambersburg PA
CBHW061248220326
41599CB00028B/5570